ELECTROCHEMICAL SENSORS TECHNOLOGY

Edited by **Mohammed Muzibur Rahman**
and **Abdullah Mohamed Asiri**

Electrochemical Sensors Technology

http://dx.doi.org/10.5772/65230
Edited by Mohammed Muzibur Rahman and Abdullah Mohamed Asiri

Contributors

Niina J Ronkainen, Stanley Okon, Igor Cretescu, Doina Lutic, Liliana Rozemarie Manea, Erica Perry Murray, Khawlah Kharashi, Kehinde Adedeji, Chang-Seop Lee, Yong Jae Kim, Xiaoxing Zhang, Ju Tang, Song Xiao, Yingang Gui, Qijin Chi, Minwei Zhang, Arnab Halder, Xianyi Cao, Chengyi Hou, Mohammed Muzibur Rahman

CBS, Edition **2018**

Published by InTech

Janeza Trdine 9, 51000 Rijeka, Croatia

Publishing Process Manager Nina Kalinic

Technical Editor SPi Global

Cover Designer InTech Design Team

Additional hard copies can be obtained from orders@intechopen.com

Electrochemical Sensors Technology, Edited by Mohammed Muzibur Rahman and Abdullah Mohamed Asiri
p. cm.
Print ISBN 978-953-51-3193-9
Online ISBN 978-953-51-3194-6

Contents

Preface

It gives us immense pleasure in introducing a book named as *Electrochemical Sensors Technology* based on the state of the art of biosensors, state of the art of gas sensors, and state of the art of chemical sensors with various nanocomposites or nanomaterials, as well as their various potential and significant sensor applications. These deal with selective and sensitive sensor aspects of the synthesis, fabrication, development, and practical sensing applications via nanocomposites or nanomaterials, self-assembly, composite particles, and hybrid polymers. The discussion of these aspects develops through the fundamentals and applied experimental routes in electrochemical approaches via the fabrication of various electroactive matrices to finally include the interfacing of the sensor science and technology world. Usually, hybrid or electroactive nanocomposite or nanomaterial certainly attains accomplishment of its classical essence and takes new directions of growth, development, continuous changes, and various sensor design applications in research and development sectors. The new route and emerging frontiers are branching out from time to time around this innovative biosensor, chemisensor, and gas sensor technology based on various layered matrices. Advances in electrochemical sensing materials with various potential instrumentations for evaluating the practical investigation of real biological/environmental samples now enable us to direct implementation that takes place on the fabricated layered with nanocomposites. This work aims to bridge the gap between graduate and researches in applied electrochemical and biological branches, as well as biomedical and environmental sensor design sciences, in order to initiate researchers into electrochemical sensor study in as straightforward way as possible and to introduce the scientist to the opportunities offered by the applied technological fields. We worked consistently to complete this work on "Electrochemical Sensors" under the InTech open access publisher. This book offers a comprehensive but easy-to-follow theoretical context with practical investigation in electrochemical science. It frameworks the significant research methods and contemplates technological implementations in real biological and environmental analyses. A straightforward acquaintance of the principles of chemical, physical, biological, and environmental sensor-designed chemistry is anticipated. It will also entreaty to a broad audience of undergraduate, graduate, and postgraduate students at colleges and universities of technology, as well as scientists in industry who prerequisite a comprehensive circumstantial study in the sensor science and technology. We hope that the book *Electrochemical Sensors Technology* would further enhance the applied electroactive nanocomposites, nanohybrides, or nanomaterials in chemical, biological, and biomedical sciences, especially in bringing new entrants into the applied electrochemical sensor science and technology fields, and help scientists to forward and develop their own field of specialization.

Prof. Mohammed Muzibur Rahman, PhD
Prof. Abdullah Mohamed Asiri, PhD
Center of Excellence for Advanced Material Research (CEAMR) and Chemistry Department,
King Abdulaziz University, Saudi Arabia

Section 1

Introduction

Chapter 1

Introductory Chapter: Electrochemical Sensors Technology

Mohammed Muzibur Rahman and
Abdullah Mohamed Asiri

Additional information is available at the end of the chapter

http://dx.doi.org/10.5772/intechopen.68709

This book describes a comprehensive overview of electro-chemical sensors and biosensors for the analyses, investigation, and monitoring of the most significant unsafe analytes in the ecological as well as environmental field in industry, in treatment plants, and in environmental research. The contributed chapters stretch the reader a comprehensive study, state-of-the-art picture of the field of electro-chemical sensors or biosensors appropriate to environmental analytes, from the theoretical principles of their design to their implementation, realization, and potential application. It covers the most recent techniques and nanocomposites/nanomaterials for the preparation, construction, validation, analyses, and design of electro-chemical sensors/biosensors for bio-analytical, clinical, and environmental applications—emphasizing the latest classes of selective, sensitive, robust, fast response, stable, electro-chemical sensors as well as electrochemical biosensors for in vivo/vitro diagnosis.

Development in advanced nanotechnology and the conservatory of innovative chemical sensors, biosensors, ionic sensors with various composites/materials and nanodevices has been a regulating key task in the fabrication and improvement of very precise, perceptive, accurate, sophisticated, sensitive, and consistent efficient chemical sensors [1–3]. The exploration for even tiny electrodes accomplished in nano-level imaging and controlling of doped nanomaterials, doping agents (host-guest), biological, chemical, pathological samples, and chemical sensors has recently extended the attention of awareness of the scientist, mainly for control monitoring, owing to the amplifying essential for environmental safety and health monitoring [4–6]. Recently, great attention is provided for the detection of various unsafe, carcinogenic, toxic, hazardous chemicals, or biomolecules to live safely and as well as to prevent the ecological system from harmful effects of toxins [7–9].

Ronkainen et al. describe the enzyme-based electrochemical glutamate biosensor development, which has been proposed to play a significant role in various neurological and psychiatric disorders. In this contribution, the design, construction, and optimization of enzyme-based principle theory. Additionally, the density of states, adsorption energy, charge transfer amount, and adsorption distance of adsorption process between CNTs and dissolved gas were also calculated in his chapter. Two kinds of sensors, mixed acid-modified CNTs and $NiCl_{2}$-modified CNTs, were prepared to conduct the dissolved gases response experiment. Afterward, the gas response mechanisms were investigated. Finally, the results between response experiment and theoretical calculation were compared, reflecting a good coherence with each other. The carbon nanotube (CNT) based gas sensors possess a relatively high sensitivity and fine linearity and could be employed in dissolved gas analysis equipment in the transformer [42–47].

Finally, this book generally reviews the recent and advanced methods and substantial applications of biosensors, gas sensor and chemical sensors. Contributed chapters are scratched by expert scientists and professors in the electrochemical sensor field. This book aims to make a connection between undergraduates, post-graduates, graduates, and scientists on their researches in sensor development based on enzyme-based sensors, graphene-based sensors, NOx sensors, gas sensors, hazardous and toxic gas sensors, and nano-sensors in environmental and biomedical sciences in order to initiate researchers into various sensors study in as straightforward a way as possible and as well as present the scientist the opportunities offered by the health care science and ecological fields. However, each chapter delivers methodological details beyond the level originally in representative journal articles and explores the potential applications of biological and chemical sensors to a substantial level in health care, real clinical, food, industrial, cancer diagnostics, biomedical, environmental science and detection of infectious organisms, also providing a brochure for the future as well as in the safety and security arena. The primary target audience for this book "**Electrochemical Sensors Technology**" includes students, researchers, technologists, physicists, chemists, biologists, engineers, and professionals who are interested in bio, chemi, and gas sensors and associated topics.

Author details

Mohammed Muzibur Rahman* and Abdullah Mohamed Asiri

*Address all correspondence to: mmrahmanh@gmail.com

Center of Excellence for Advanced Materials Research (CEAMR) & Chemistry Department, King Abdulaziz University, Jeddah, Saudi Arabia

References

[1] Alam MK, Rahman MM, Abbas M, Torati SR, Asiri AM, Kim D, Kim CG. Ultra-sensitive 2-nitrophenol detection based on reduced graphene oxide/ZnO nanocomposites. Journal of Electroanalytical Chemistry. 2017;**788**:66-73

[2] Hussain MM, Rahman MM, Arshad MN, Asiri AM. Hg^{2+} Sensor Development Based on (*E*)-N′-Nitrobenzylidene-Benzenesulfonohydrazide (NBBSH) Derivatives Fabricated on a Glassy Carbon Electrode with a Nafion Matrix. ACS Omega. 2017;**2**:420-431

Chapter 1

Introductory Chapter: Electrochemical Sensors Technology

Mohammed Muzibur Rahman and
Abdullah Mohamed Asiri

Additional information is available at the end of the chapter

http://dx.doi.org/10.5772/intechopen.68709

This book describes a comprehensive overview of electro-chemical sensors and biosensors for the analyses, investigation, and monitoring of the most significant unsafe analytes in the ecological as well as environmental field in industry, in treatment plants, and in environmental research. The contributed chapters stretch the reader a comprehensive study, state-of-the-art picture of the field of electro-chemical sensors or biosensors appropriate to environmental analytes, from the theoretical principles of their design to their implementation, realization, and potential application. It covers the most recent techniques and nanocomposites/nanomaterials for the preparation, construction, validation, analyses, and design of electro-chemical sensors/biosensors for bio-analytical, clinical, and environmental applications—emphasizing the latest classes of selective, sensitive, robust, fast response, stable, electro-chemical sensors as well as electrochemical biosensors for in vivo/vitro diagnosis.

Development in advanced nanotechnology and the conservatory of innovative chemical sensors, biosensors, ionic sensors with various composites/materials and nanodevices has been a regulating key task in the fabrication and improvement of very precise, perceptive, accurate, sophisticated, sensitive, and consistent efficient chemical sensors [1–3]. The exploration for even tiny electrodes accomplished in nano-level imaging and controlling of doped nanomaterials, doping agents (host-guest), biological, chemical, pathological samples, and chemical sensors has recently extended the attention of awareness of the scientist, mainly for control monitoring, owing to the amplifying essential for environmental safety and health monitoring [4–6]. Recently, great attention is provided for the detection of various unsafe, carcinogenic, toxic, hazardous chemicals, or biomolecules to live safely and as well as to prevent the ecological system from harmful effects of toxins [7–9].

Ronkainen et al. describe the enzyme-based electrochemical glutamate biosensor development, which has been proposed to play a significant role in various neurological and psychiatric disorders. In this contribution, the design, construction, and optimization of enzyme-based electrochemical biosensors for in vivo and in vitro detection of glutamate were discussed in this contribution [10–17]. Various glutamate biosensors have been discussed, including the developed glutamate monitoring dynamic levels of extracellular glutamate in the living brain tissue adding to the current medical knowledge of these complex neurotransmitter systems and ultimately impacting treatment plans [18–21]. More significantly, glutamate biosensors have been used in environmental monitoring, in the fermentation industry, and in the food industry for determination of Monosodium glutamate (MSG), a common flavor-enhancing food additive. With continuous developments in molecular biology, nanofabrication methods, immobilization methods of biomolecules and multiplexing capabilities, the production of sensitive, selective, fast, and easy-to-use biosensors for quantification of glutamate, and other neurotransmitters will be feasible in the not too distant future.

Qijin et al. approached some graphene paper-based electro-chemical sensors to illustrate recent advances in the research and development of 2D graphene papers as new and noble materials for electro-chemical sensors. It covers the design, fabrication, functionalization, and application evaluation of graphene papers. Precise monitoring of chemical or biological processes is of extreme importance for medical and biological applications. Electro-chemical sensors can ideally fulfill that goal by converting a chemical or biological response into a processable and quantifiable signal. In the past two decades, intensive research and development of electro-chemical sensors have enabled to fabricate different types of devices [22–31]. After the development of many successful commercial electro-chemical sensors in the classic configurations, currently, there is a notable transition and increasing demands for the development of flexible and wearable sensors. The development of flexible electro-chemical sensors depends crucially on the discovery and preparation of freestanding and flexible new materials. They first summarized the mainstream methods for fabrication of graphene papers/membranes with the focus on chemical vapor deposition techniques and solution-processing assembly. A large portion of this work is devoted to the highlights of specific functionalization of graphene papers with polymer and nanoscale functional building blocks for electrochemical sensing purposes. In terms of electrochemical sensing applications, the emphasis is on enzyme-graphene and nanoparticle-graphene paper-based systems for detection of glucose. We conclude this chapter with brief remarks and an outlook. In short, worldwide researchers have explored graphene paper–based sensors by exploiting their unique advantages including high sensitivity, conductivity, and in-situ sensing. The recent research advances suggest that graphene paper-based materials could play a significant role in developing flexible sensors and electronic devices due to their intriguing structural and functional features.

Murray et al. focused on managing H_2O cross-sensitivity using composite electrolyte "NO_x sensors". They approached NO_x sensors composed of PSZ, FSZ, and PSZ–FSZ composite electrolytes which were investigated using impedance spectroscopy under dry and humidified gas conditions. The microstructural properties, NO_x sensitivity, oxygen partial pressure and temperature dependence, as well as the response time of the sensors composed of the various electrolytes were characterized in order to interpret the electrochemical response with respect to water cross-sensitivity. In this approach, impedance spectroscopy was used to interpret the electrochemical response of NO_x sensors composed of PSZ, FSZ, and PSZ–FSZ composite electrolytes during operation under dry and humidified gas conditions. Analysis of the electrochemical responses of the 50PSZ–50FSZ-based sensors indicated PSZ contributed to lower water cross-sensitivity, while FSZ promoted NO_x sensitivity. Finally, sensors composed of the 50PSZ–50FSZ composite electrolyte demonstrated significant sensitivity to NO and low cross-sensitivity to water with negligible temperature dependence [32–35].

Lutic et al. describe electro-chemical sensors for monitoring indoor and outdoor air pollution. They approached a comprehensive presentation of the most common electro-chemical sensors used in real monitoring applications of air purity testing. The air quality monitoring stations based on electro-chemical sensors are nowadays used to determine the global pollution index of the atmospheric air, in order to prevent the risks toward human health and damage of environment, especially in the highly populated and industrialized urban areas. The electro-chemical gas sensors are nowadays indispensable in the monitoring of the atmosphere quality, especially due to pollutants associated with human activities. Carbon monoxide, sulfur oxides, hydrogen sulfide, and nitrogen oxides are only a few of species which can seriously damage the environment equilibrium by smog formation, acid rain, soil deterioration, water contamination as well as some direct damages on the human health [36–38]. The electrochemical gas sensing is based on gas oxidation or reducing reactions on sensing surfaces with catalytic potential, surfaces which suffer noticeable charge changes, that can be amplified and processed in order to generate a signal. The electro-chemical sensors are fast, reliable, small and cheap; therefore, their use covers nowadays the exhaust systems from automotives, domestic/residential gas detection, and leak checkers. Respecting the rated voltage as said by the manufacturer, using the sensor in the right temperature range, avoiding the deterioration due to exposure to humidity, avoiding contamination with various chemicals, lack of sudden exposure to extreme temperatures, and avoiding the mechanical shocks are basic conditions to preserve their work function and accuracy.

Lee et al. have focused on fabrication and characteristics of metal-loaded mixed metal oxides gas sensors for the detection of toxic gases for environmental purposes. They approached developing gas sensors which permit individuals to circumvent poisonous gases that may be produced in spaces with residues of inorganic/organic waste with certain temperature at 50°C or above. The response, sensitivity, and selectivity of these gas sensors to types of carcinogenic gases such as H_2S, toluene, and aldehyde were examined. The thick-film semiconductor sensors that detect some toxic gases were fabricated using nano-sized sensing materials powder (SnO_2, WO_3, and ZnO), and these were prepared via sol-gel and precipitation methods. Response to various lethal gases was measured and is defined as the ratio (Ra/Rg) of the resistance of the sensor film in air to the resistance of the film in toxic gas. Generally, semiconductor metal-oxide gas sensors can be used for diverse applications, ranging from equipment to monitor environmental and occupational safety to facilitating quality assurance through novel measurement. The nature of the gas-sensitive material and the concentration of the target gas (usually a few ppb~ppm) determine the measuring range and limitations of the device [39–41].

Zhang describes the potential application of nanosensors in dissolved gases for the detection in oil-insulated transformers in this contribution. Here, it is approached on the adsorption processes between modified CNTs (CNTs-OH, Ni-CNTs) and dissolved gases in transformers oil including C_2H_2, C_2H_4, C_2H_6, CH_4, CO, and H_2 which have been simulated based on the first principle theory. Additionally, the density of states, adsorption energy, charge transfer amount, and adsorption distance of adsorption process between CNTs and dissolved gas were also calculated in his chapter. Two kinds of sensors, mixed acid-modified CNTs and $NiCl_2$-modified CNTs, were prepared to conduct the dissolved gases response experiment. Afterward, the gas response mechanisms were investigated. Finally, the results between response experiment and theoretical calculation were compared, reflecting a good coherence with each other. The carbon nanotube (CNT) based gas sensors possess a relatively high sensitivity and fine linearity and could be employed in dissolved gas analysis equipment in the transformer [42–47].

Finally, this book generally reviews the recent and advanced methods and substantial applications of biosensors, gas sensor and chemical sensors. Contributed chapters are scratched by expert scientists and professors in the electrochemical sensor field. This book aims to make a connection between undergraduates, post-graduates, graduates, and scientists on their researches in sensor development based on enzyme-based sensors, graphene-based sensors, NOx sensors, gas sensors, hazardous and toxic gas sensors, and nano-sensors in environmental and biomedical sciences in order to initiate researchers into various sensors study in as straightforward a way as possible and as well as present the scientist the opportunities offered by the health care science and ecological fields. However, each chapter delivers methodological details beyond the level originally in representative journal articles and explores the potential applications of biological and chemical sensors to a substantial level in health care, real clinical, food, industrial, cancer diagnostics, biomedical, environmental science and detection of infectious organisms, also providing a brochure for the future as well as in the safety and security arena. The primary target audience for this book "**Electrochemical Sensors Technology**" includes students, researchers, technologists, physicists, chemists, biologists, engineers, and professionals who are interested in bio, chemi, and gas sensors and associated topics.

Author details

Mohammed Muzibur Rahman* and Abdullah Mohamed Asiri

*Address all correspondence to: mmrahmanh@gmail.com

Center of Excellence for Advanced Materials Research (CEAMR) & Chemistry Department, King Abdulaziz University, Jeddah, Saudi Arabia

References

[1] Alam MK, Rahman MM, Abbas M, Torati SR, Asiri AM, Kim D, Kim CG. Ultra-sensitive 2-nitrophenol detection based on reduced graphene oxide/ZnO nanocomposites. Journal of Electroanalytical Chemistry. 2017;**788**:66-73

[2] Hussain MM, Rahman MM, Arshad MN, Asiri AM. Hg^{2+} Sensor Development Based on (*E*)-N′-Nitrobenzylidene-Benzenesulfonohydrazide (NBBSH) Derivatives Fabricated on a Glassy Carbon Electrode with a Nafion Matrix. ACS Omega. 2017;**2**:420-431

[3] Rahman MM, Jamal A, Khan SB, Faisal M. Cu-doped ZnO based nanostructured materials for sensitive chemical sensor applications. ACS Applied Material and Interfaces. 2011;**3**:1346-1351

[4] Rahman MM, Balkhoyor HB, Asiri AM. Phenolic sensor development based on chromium oxide-decorated carbon nanotubes for environmental safety. Journal of Environmental Management. 2017;**188**:228-237

[5] Rahman, A Jamal, SB Khan, M Faisal. Fabrication of highly sensitive ethanol chemical sensor based on Sm-Doped Co_3O_4 nano-kernel by solution method. Journal of Physical Chemistry. 2011;**C 115**:9503-9510

[6] Rahman MM, A Jamal, SB Khan, M Faisal. Highly sensitive ethanol chemical sensor based on Ni-doped SnO_2 nanostructure materials. Biosensors and Bioelectronics. 2011;**28**: 127-134

[7] Umar A, Rahman MM, Kim SH, Hahn YB. Zinc oxide nanonail based chemical sensor for hydrazine detection. Chemical Communication. 2008:166-169. DOI:10.1039/B711215G

[8] Rahman MM, Hussain MM, Asiri AM. A novel approach towards the hydrazine sensor development by SrO.CNT nanocomposites. RSC Advances.2016;**6**:65338-65348

[9] Rahman MM, Abu-Zied BM, Hasan MM, Asiri AM, Hasnat MA. Fabrication of selective 4-aminophenol sensor based on H-ZSM-5 zeolites deposited silver electrodes. RSC Advances. 2016;**6**:48435-48444

[10] Black DW, Andreasen NC. Introductory Textbook of Psychiatry. 6th ed. Arlington: American Psychiatric Publishing; 2014

[11] Ronkainen NJ, Okon SL. Nanomaterial-based electrochemical immunosensors for clinically significant biomarkers. Materials. 2014;**7**:4669-4709

[12] Hussain MM, Rahman MM, Asiri AM. Sensitive L-Leucine sensor based on a glassy carbon electrode modified with SrO nanorods. Microchimica Acta. 2016;**183**:3265-3273

[13] Rahman MM. Fabrication of mediator-free glutamate sensors based on glutamate oxidase using smart micro-devices. Journal of Biomedical Nanotechnology. 2011;**7**:351-357

[14] Rahman MM, Asiri AM. Development of Penicillin G biosensor based on Penicillinase enzymes immobilized biochips. Biomedical Microdevices. 2015;**17**:9

[15] Rahman MM, Asiri AM. One-step electrochemical detection of cholesterol in presence of suitable $K_3Fe(CN)_6$/phosphate buffer mediator by an electrochemical approach. Talanta. 2015;**140**:96-101

[16] Rahman MM, Asiri AM. Selective choline biosensors based on choline oxidase co-immobilized into self-assembled monolayers on micro-chips at low potential. Analytical Methods. 2015;**7**:9426-9434

[17] Hussain MM, Rahman MM, Asiri AM, Awual MR. Non-enzymatic simultaneous detection of L-glutamic acid and uric acid using mesoporous Co_3O_4 nanosheets. RSC Advances. 2016;**6**:80511-80521

[18] Rahman MM, Hussain MM, Asiri AM. A glutathione biosensor based on a glassy carbon electrode modified with CdO nanoparticles-decorated carbon nanotube in a nafion matrix. Microchimica Acta. 2016;**183**:3255-3263

[19] Field JR, Walker AG, Conn PJ. Targeting glutamate synapses in schizophrenia). Trends in Molecular Medicines. 2011;**17**:689-698

[20] Rahman MM, Ahmed J, Asiri AM. A glassy carbon electrode modified with γ-Ce2S3-decorated CNT nanocomposites for uric acid sensor development: A real sample analysis. RSC Advances. 2017;**7**:14649-14659

[21] Ronkainen NJ, Halsall HB, Heineman WR. Electrochemical biosensors. Chemical Society Reviews. 2010;**39**:1747-1763

[22] Wang J, Musameh M. Carbon nanotube/Teflon composite electrochemical sensors and biosensors. Analytical Chemistry. 2003;**75**:2075-2079

[23] Halder A, Zhang M, Chi Q. Electroactive and biocompatible functionalization of graphene for the development of biosensing platforms. Biosensors And Bioelectronics. 2017;**87**:764-771

[24] Rahman MM, Ahmed J, Asiri AM. Development of creatine sensor based on antimony-doped Tin oxide (ATO) nanoparticles. Sensors and Actuators B: Chemical. 2017;**242**: 167-175

[25] Zhu N, Han S, Gan S, Ulstrup J, Chi Q. Graphene paper doped with chemically compatible Prussian Blue nanoparticles as nanohybrid electrocatalyst. Advanced Functional Materials. 2013;**23**:5297-5306

[26] Zhang M, Hou C, Halder A, Ulstrup J, Chi Q. Interlocked graphene Prussian blue hybrid composites enable multifunctional electrochemical applications. Biosensors And Bioelectronics. 2017;**89**:570-577

[27] Lee SR,Rahman MM, Ishida M, Sawada K. Fabrication of a highly sensitive penicillin sensor based on charge transfer techniques. Biosensensors and Bioelectronics. 2009;**24**:1877-1882

[28] Zhang M, Hou C, Halder A, Chi Q. Ultralight, flexible and semi-transparent metal oxide papers for photoelectrochemical water splitting. ACS Applied Materials And Interfaces. 2017;**9**:3922-3930

[29] Zhang M, Halder A, Hou C, Ulstrup J, Chi Q. Free-standing and flexible graphene papers as disposable non-enzymatic electrochemical sensors. Bioelectrochemistry. 2016;**109**:87-94

[30] Lee SR, Rahman M, Ishida M, Sawada K. Development of highly sensitive acetylcholine sensor based on acetylcholine by charge transfer techniques esterase using smart biochips. Trends in Analytical Chemistry. 2009;**28**:196-203

[31] Zhang M, Hou C, Halder A, Wang H, Chi Q. Graphene papers: Smart architecture and specific functionalization for biomimetics, electrocatalytic sensing and energy storage. Materials Chemistry Frontiers. 2017;**1**:37-60

[32] Kharashi K, Murray EP. Effect of Al_2O_3 in porous zirconia electrolytes for NO sensing. Journal of The Electrochemical Society. 2016;**163**:B633-B637

[33] Martin LP, Woo LY, Glass RS. Impedancemetric NO_x sensing using YSZ electrolyte and YSZ/Cr_2O_3 composite electrodes. Journal of The Electrochemical Society. 2007;**154**: J97-J104

[34] Hussain MM, Rahman MM, Asiri AM. Efficient 2-nitrophenol chemical sensor development based on Ce_2O_3 nanoparticles decorated CNT nanocomposites for environmental safety. PLoS ONE. 2016;**11**:e0166265

[35] Woo L, Martin LP, Glass RS, Wensheng W, Sukwon J, Gorte RJ, Murray EP, Novak RF, Visser JH. Effect of electrode composition and microstructure on impedancemetric nitric oxide sensors based on YSZ electrolyte. Journal of the Electrochemical Society. 2008;**155**:J32-J40

[36] Arshak K, Moore E, Lyons GM, Harris J, Clifford S. A review of gas sensors employed in electronic nose applications. Sensor Review. 2004;**24**:181-198

[37] Rahman MM, Balkhoyor HB, Asiri AM. Ultrasensitive and selective hydrazine sensor development based on Sn/ZnO nanoparticles. RSC Advances. 2016;**6**:29342-29352

[38] Zhuiykov S, Miura N. Development of zirconia-based potentiometric NOx sensors for automotive and energy industries in the early 21st century: What are the prospects for sensors?Sensors and Actuators B. 2007;**212**:639-651

[39] Noordally E, Richmond JR, Tahir SF. Destruction of volatile organic compounds by catalytic oxidation. Catalysis Today. 1993;**17**:359-366

[40] Rahman MM, Ahmed J, Asiri AM, Siddiquey IA, Hasnat MA. Development of ultrasensitive hydrazine sensor based on facile CoS_2-CNT nanocomposites. RSC Advances. 2016;**6**:90470-90479

[41] Hodgson AT, Faulkner D, Sullivan DP, DiBartolomeo DL, Russell ML, Fisk WJ. Effect of outside air ventilation rate on volatile organic compound concentrations in a call center. Atmospheric Environment. 2003;**37**:5517-5527

[42] Zhang X, Gui Y, Dai Z. A simulation of Pd-doped SWCNTs used to detect SF_6 decomposition components under partial discharge. Applied Surface Science. 2014;**315**:196-202

[43] Zhang X, Chen Q, Tang J, Hu W, Zhang J. Adsorption of SF_6 decomposed gas on anatase (101) and (001) surfaces with oxygen defect: A density functional theory study. Scientific Reports. 2014;**4**:4762

[44] Rahman MM, Jamal A, Khan SB, Faisal M, Asiri AM. Fabrication of highly sensitive acetone sensor based on sonochemically prepared as-grown Ag_2O nanostructures. Chemical Engineering Journal. 2012;**192**:122-128

[45] Zhang X, Tie J, Zhang J. A Pt-doped TiO_2 nanotube arrays sensor for detecting SF_6 decomposition products. Sensors. 2013;**13**:14764

[46] Ebbesen TW, Ajayan PM. Large-scale synthesis of carbon nanotubes. Nature. 1990;**358**: 220-222

[47] Zhang X, Gui Y, Dong X. Preparation and application of TiO_2 nanotube array gas sensor for SF_6-insulated equipment detection: A review. Nanoscale Research Letters. 2016;**11**: 1-13

Section 2

State-of-the-Art Biosensors

Chapter 2

Enzyme-Based Electrochemical Glutamate Biosensors

Stanley L. Okon and Niina J. Ronkainen

Additional information is available at the end of the chapter

http://dx.doi.org/10.5772/68025

Abstract

Glutamate, a major excitatory neurotransmitter in the mammalian central nervous system, plays a vital role in many physiological processes and is one of the key neurotransmitters of interest in psychopharmacology. It is involved in many normal and abnormal behaviors related to neurological and psychiatric disorders. The glutamate system has been proposed to play a significant role in various neurological and psychiatric disorders such as Alzheimer's disease, autism, schizophrenia, depression, drug addiction, and more. The design, construction, and optimization of enzyme-based electrochemical biosensors for in vivo and in vitro detection of glutamate are active areas of interdisciplinary research. For example, various glutamate biosensors have been developed for monitoring dynamic levels of extracellular glutamate in the living brain tissue adding to the current medical knowledge of these complex neurotransmitter systems and ultimately impacting treatment plans. In addition to biological sciences and clinical medicine, glutamate biosensors have been used in environmental monitoring, in the fermentation industry, and in the food industry for determination of monosodium glutamate (MSG), a common flavor-enhancing food additive.

Keywords: biocatalytic sensors, L-glutamate, mediator, monosodium glutamate, carbon nanotubes, nanomaterials, neurological and psychiatric disorders, glutamate oxidase, glutamate dehydrogenase

1. Introduction

Glutamate is a nonessential amino acid, a precursor for gamma aminobutyric acid (GABA (the primary inhibitory neurotransmitter)), and an abundant excitatory neurotransmitter in the mammalian central nervous system that is produced by pyramidal cells located in the cerebral cortex and hippocampus. It has an important role in the formation and stabilization of synapses, long-term potentiation (i.e., the long-lasting enhancement in signal transmission between two neurons which results from synchronously stimulating them), neurodegenerative diseases, learning, and the formation of memories. In addition, maintaining a balance in glutamate levels in the central nervous system is important because increased levels lead to neurotoxicity and cell death, while decreased levels result in impaired long-term potentiation, impaired synaptic plasticity (i.e., the ability of synapses to change their structures in response to inputs and changes in their environment), and impaired cognitive performance [1]. Glutamate also plays a pivotal role in cellular metabolism as it is associated with transamination reaction, a key step in amino acid degradation, and is the product of deamination. Hence, quantifying glutamate levels in biological fluids and tissues reliably and reproducibly is of interest in many disciplines.

Glutamate dysregulation may induce excitotoxicity, which is closely associated with multiple psychiatric and cognitive disorders. Glutamate is hypothesized to have a key role in the pathophysiology of psychiatric disorders such as schizophrenia [2] and depression [3]. It is also currently a target of novel drugs for potential treatments of schizophrenia and depression. Briefly, the glutamate hypothesis of schizophrenia suggests that there is a hypofunction in *N*-methyl-D-Aspartate (NMDA)-type glutamate receptors within the glutamate system wherein the normal balance between excitatory glutamate and inhibitory GABA is destabilized resulting in a combination of excitotoxicity and impaired neuroplasticity, which results in the psychotic symptoms of schizophrenia. As stated earlier, glutamate or glutamic acid is an excitatory amino acid neurotransmitter that is synthesized from the precursor amino acid glutamine, in cells called glia which protect and support neurons in the nervous system. In addition, glia recycle and regenerate glutamate after its release during neurotransmission. After glutamate is released from glutamate neurons, it binds to synaptic receptors and is then pumped into glia by an excitatory amino acid transporter (EAAT). Once in the glia, glutamate is converted into glutamine by the enzyme glutamate synthetase. Glutamine is released from glia via reverse transport either by the specific neutral amino acid transporter (SNAT) or by the alanine-serine-cysteine transporter (ASC-T) and transported into the neuron. Inside the neuron, glutamine is converted to glutamate by a mitochondrial enzyme called glutaminase for use as a neurotransmitter. After use, glutamate is stored in synaptic vesicles for subsequent release during neurotransmission, and its actions are terminated via removal by EAATs [4]. The next section will outline the proposed role of glutamate in schizophrenia.

2. Glutamate and schizophrenia

In the Diagnostic and Statistical Manual of Mental Disorders, Fifth Edition (DSM-5), schizophrenia is defined by a group of characteristic symptoms which consist of (1) psychotic or positive symptoms (i.e., symptoms characterized by the presence of something that should be absent, such as hallucinations or delusions), (2) disorganized symptoms (such as disorganized speech–e.g., incoherence or frequent derailment, disorganized or catatonic behavior, and inappropriate affect), (3) negative symptoms (i.e., symptoms characterized by the absence of something that should be present, such as avolition [lack of motivation], affective flattening [diminished emotional expression], or alogia [poverty of speech]), (4) deterioration in social, occupational, or interpersonal relationships, and (5) continuous signs of the disturbance for at least 6 months [1].

Glutamate-based theories of schizophrenia resulted from the observation that two related compounds—phencyclidine (PCP) and ketamine—produced symptoms and cognitive deficits in healthy subjects which resembled those found in schizophrenia. These symptoms, including negative as well as positive symptoms, resolved following elimination of the compounds. Years later, further research revealed that these compounds produced their effects by blocking neurotransmission at NMDA-type glutamate receptors which in turn gave rise to current glutamate-based models of schizophrenia. In addition, it has been found that many suspected genes for schizophrenia affect glutamate neurotransmission. Also, NMDA receptors are regulated by brain levels of glycine, D-serine, and glutathione such that disturbances in concentrations of these compounds and of the genes related to their synthesis may all contribute to schizophrenia. Since available data to date suggests that there is more than one single cause of NMDA receptor dysfunction among different individuals, perhaps threshold models are more useful in that for each individual, the sum of genetic and environmental factors determines whether NMDA receptor functions will fall below a critical level. In schizophrenia, it is postulated that psychosis may emerge once the level of NMDA receptor function decreases by approximately 20%, with worsening symptom severity thereafter. Furthermore, since glutamate and NMDA receptors are widely distributed throughout the brain, this has led to a concept of "whole brain" dysfunction, which involves the prefrontal, limbic, auditory, and visual cortices among others. Some specific examples of the consequences of glutamate deficits in certain brain regions include (1) working memory impairments and inability to unlearn dysfunctional behavior patterns due to glutamate deficits in the dorsolateral prefrontal cortex, (2) impaired response inhibition and hence an increased tendency for impulsivity and impulsive aggression due to glutamate deficits in the inferior prefrontal cortex, (3) impairments in learning and memory formation which contribute to psychosis and delusion formation due to glutamate deficits in the medial temporal cortex, (4) perceptual changes including impaired reading ability due to glutamate deficits in the visual cortex, and (5) impaired ability to detect vocal intonation which leads to impairments in emotional recognition and social cognition due to glutamate deficits in the primary auditory cortex. Thus, widespread cognitive dysfunction in these brain regions due to dysfunctional glutamate and NMDA-type glutamate receptors represents a formidable obstacle that prevents individuals with schizophrenia from returning to premorbid levels of functioning [5].

3. Other applications of glutamate biosensors

In addition to its key roles in fundamental neurological processes and neurological disorders, glutamate is important in protein synthesis, protein degradation, and nitrogen metabolism. The concentrations of glutamate in intracellular environment vary from 2 to 20 mM [6]. Usual glutamate concentrations in plasma are approximately 150 and 10 μM in cerebrospinal fluid [7, 8]. Normal glutamate concentration in the extracellular space ranges between 1 and 80 μM [9]. The quantification of L-glutamate is also important in food analysis due to questions about its safety as a food additive. Monosodium glutamate (MSG, $C_5H_8NO_4Na$), invented by Dr. Kikunae Ikeda in 1908, is a commonly used flavor-enhancing additive found in Chinese restaurant food [10, 11], canned soups, canned vegetables, and processed meats. Interestingly, there has been controversy surrounding the use and safety of MSG as a food additive. According to some studies, excessive intake of MSG may cause headaches and stomach pain in certain individuals as well as neuronal excitotoxicity. However, use of MSG as a food additive is generally regarded as harmless. Still, many food manufacturers choose to advertise their products as being MSG-free. Development of MSG biosensors for food applications is also an active area of research. For example, Monošík et al. recently prepared and characterized a bienzymatic nanocomposite electrode for quantification of MSG in food samples, utilizing L-glutamate dehydrogenase and diaphorase enzymes immobilized between chitosan layers on nanocomposite electrodes consisting of multiwalled carbon nanotubes (MWCNTs) [12]. The structure of monosodium glutamate is shown in **Figure 1**.

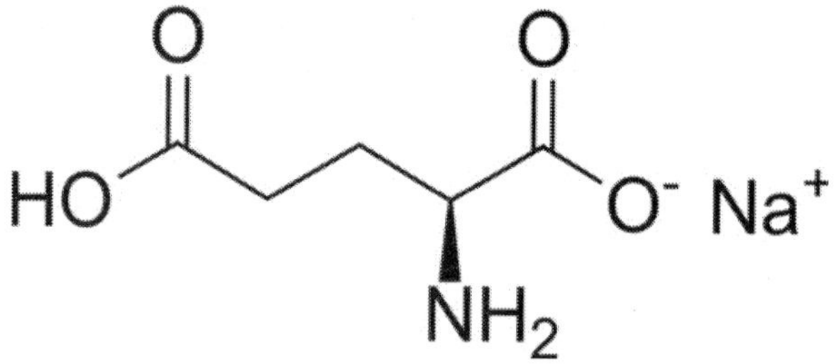

Figure 1. The chemical structure of monosodium glutamate.

4. Glutamate biosensors

This chapter will primarily focus on providing an overview about various enzyme-based glutamate biosensors that utilize sensitive electrochemical detection methods. In addition, the characterization and optimization of newly developed enzyme-based electrochemical biosensors will be discussed briefly. A brief overview of the most common electrochemical detection methods utilized in biocatalytic glutamate sensor characterization, testing, and in quantitative analysis will also be provided. We will begin with a brief introduction to electrochemical biocatalytic sensors.

In general terms, biosensors are devices that register a biochemical reaction which is then converted into a signal that can be detected and quantified [13]. A typical biosensor contains biological recognition molecules, such as enzymes, that are highly selective and specific for a given analyte. In electrochemical biocatalytic sensors, the biological recognition molecules bind reversibly to a particular analyte on or near an electrochemically active interface which may incorporate nanomaterials, giving rise to a measurable signal [13]. An electric transducer, usually a modified electrode, which is in contact with the electrochemically active interface, converts the biochemical reaction into an electrical signal that is further amplified by a signal processor into a useful form. The development of electrochemical biosensors for determination of glutamate is an active area of research as these sensors are typically easy to use, fast, reliable, convenient, portable, and affordable. Various instrumental analysis techniques such as spectrophotometry [14, 15], chemiluminescence [16], capillary electrophoresis [17–19], gas chromatography [20], and high-performance liquid chromatography [21–23] have also been utilized in detecting glutamate. However, these methods can be expensive, time-consuming, labor intensive, often require sample preparation, and utilize sophisticated instruments which require trained personnel to operate. Specifically, chromatography methods often require analyte derivatization, while spectrophotometry requires tedious sample pretreatment procedures.

Glutamate biosensors have been developed for both in vitro and in vivo applications. One of the challenges surrounding the utilization of sensitive and selective in vivo monitoring of dynamic levels of extracellular glutamate in living tissues using biosensors is that unlike an electroactive analyte such as dopamine, glutamate is nonelectroactive. Therefore, the direct measurement of glutamate using voltammetric electroanalytical techniques is not possible. However, if a biological recognition component such as an enzyme is added onto a physical transducer, in this case the electrode, the glutamate levels may be measured indirectly by quantifying one of the enzymatically generated products at the biocatalytic sensor [13]. A few commercial biosensors for routine glutamate measurements in the food industry are available from companies such as Yellow Springs Instruments (Yellow Springs, OH, USA).

Microdialysis methods have been widely used in monitoring extracellular glutamate levels in the brains of conscious animals. Yao and Okano described an in vivo flow-injection biosensor system with online microdialysis samples attempting to simultaneously determine the concentrations of L-glutamate, dopamine, and acetylcholine [24]. The triple electrode measured an average of 6 μM of L-glutamate in rat brain, whereas dopamine and acetylcholine levels were below the detection limit of the biosensor.

Furthermore, studies have shown that the glutamate detected using microdialysis is not or is only partly derived from synaptic transmission. As a result, development of appropriate biosensors for the accurate detection and monitoring of glutamate in vivo remains an ongoing endeavor. The ideal in vivo biosensors would be enzyme-based electrochemical biosensors in which a highly specific enzyme is immobilized on the surface of a sensitive electrochemical transducer. Indeed, certain authors have reported the fabrication and utilization of glutamate biosensors capable of second-by-second in vivo monitoring of glutamate levels in freely moving rats and mice [25, 26].

Nanotechnology and advances in microfabrication technology have played a crucial role in the fabrication of many biosensors. Nanoscale materials, with at least one dimension ranging in size from 10^{-7} to 10^{-9} m, have been incorporated into various enzymatic biosensors including many of the more recent L-glutamate sensors. The trend to manufacture smaller and more portable biosensor devices with improved performance has in part led to the incorporation of nanomaterials into biosensors. The unique chemical and physical properties of the nanomaterials also enhance the analytical performance of these biosensors. The nanomaterials utilized include carbon nanotubes (CNTs), graphene, nanowires, carbon fibers, and metal nanoparticles (NPs). Electrochemical biosensors that incorporate nanomaterials such as CNTs on the transducer surface generally have very good conductivities and much greater surface areas onto which enzymes and other molecules may be immobilized.

Recently, combining the high bioselectivity and specificity of oxidoreductase enzymes with the numerous and advantageous chemical and physical properties of organic and inorganic nanomaterials such as graphene, carbon nanotubes, carbon fibers, and gold has resulted in the development of highly sensitive and stable electrochemical biosensors with significantly improved performance for many analytes of medical interest, including glutamate.

5. Glutamate oxidase–based biosensors

Many glutamate biosensors are based on quantifying the oxidation of hydrogen peroxide (H_2O_2) liberated in a chemical reaction between glutamate oxidase (GmOx, an oxidoreductase enzyme, EC 1.4.3.11) and L-glutamate in the presence of oxygen (O_2), water (H_2O), and a flavin adenine dinucleotide (FAD) cofactor [27–45]. GmOx enzyme catalyzes the oxidative deamination of glutamate resulting in the formation of 2-oxoglutarate (i.e., α-ketoglutarate, α-KG), ammonia (NH_3), and H_2O_2 [46].

$$\text{L-glutamate} + O_2 + H_2O \rightarrow \text{2-oxoglutarate} + NH_3 + H_2O_2 \tag{1}$$

The oxidation of hydrogen peroxide occurs at the electrode surface or at very short distances away within the sample solution. The resulting current (i.e., detectable signal) which is proportional to the concentration of redox active species is then quantified by the transducer.

GmOx is a highly selective enzyme [47] and therefore unlike for many catalytic enzyme-based biosensors, interference from unwanted enzymatic reactions with similar molecular species is not a major concern for these glutamate biosensors. However, there is one study which mentions that a GmOx enzyme had a slight sensitivity (0.6 %) for L-aspartate, an amino acid with a similar side-chain group to L-glutamate [46].

In addition, electrooxidation of the GmOx-generated hydrogen peroxide requires relatively high positive potential at which common electroactive interferents, such as ascorbic acid and dopamine, also undergo oxidation thereby adding to the current (i.e., the signal) [48]. Thus, the elimination of interference in glutamate biosensors that utilize GmOx as the biorecognition molecule is critical. Strategies for the elimination of this interference by other electroactive species (which are commonly found in the sample matrix) include coating the biosensor with permselective nonconductive or conductive polymers such as Nafion, cellulose acetate, o-polyphenylenediamine (PPD), polypolyaniline, polythiophene, or polypyrrole [9, 48, 49]. The idea behind the addition of a permselective membrane is that the small pores within the membrane or film will only allow certain small molecules to pass through therefore minimizing interference by other larger electroactive species (see **Figure 2**). Also, positively

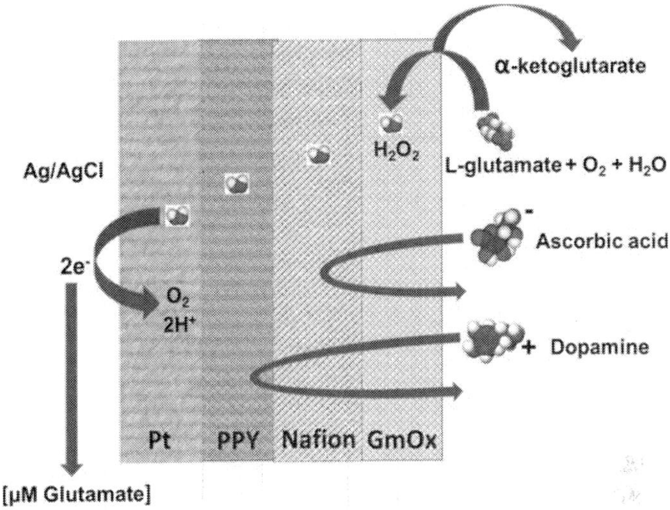

Figure 2. Elimination of interference in glutamate biosensors based on GmOx enzyme.

charged groups such as those found in sulfonated tetrafluoroethylene copolymer (Nafion) will prevent or minimize the diffusion of anionic sample components, such as ascorbic acid, across the membrane and onto the electrode surface where the redox reactions are detected. Furthermore, though membranes and composites incorporating permselective conducting polymers such as polypyrrole (PPY) have other desirable properties (such as high conductivity and being redox active), it should be noted that coating the biosensor transducer with these polymeric films can lead to longer response times and lower signals due to the added diffusion barrier for both the substrate for the enzyme and the redox active species produced in the enzyme-catalyzed reaction. Co-immobilization of peroxidase with a redox polymer [50], immobilization of ascorbate oxidase [9], and self-referencing [51] have also been utilized as strategies to minimize interference from other electroactive species. On the other hand, performing electrochemical peroxidation, a process that utilizes sacrificial electrodes and stoichiometrically balanced applications of hydrogen peroxide to efficiently destroy interfering species in the aqueous phase, poses a risk of also oxidizing the analyte of interest.

Many glutamate biosensors based on the GmOx-catalyzed reaction also include redox mediators or a second enzyme that reacts with H_2O_2 and are often referred to as second-generation biosensors. By incorporating a redox mediator, it is possible to lower the potential required for the H_2O_2 oxidation, thereby further limiting interference in complex biological samples by other species present, such as ascorbic acid or uric acid, which may also be redox active at the higher detection potential. Commonly used redox mediators in enzymatic biosensors include ferrocyanide [12], ferrocene and ferrocene derivatives [52–54], osmium complexes [55], quinine derivatives, and hexacyanoferrates, such as Prussian blue [45, 56, 57] and Ruthenium purple [58].

Prussian blue is the oldest coordination compound and was serendipitously discovered by Diesbach, an artist in 1704. It possesses excellent electrochemical characteristics and exceptional catalytic properties, making it a popular redox mediator. It is often referred to as the "artificial peroxidase" and has been well characterized and incorporated into various high-performing enzyme-based biosensors due to its excellent electrocatalysis toward the reduction of enzyme-generated H_2O_2 [45, 57, 59, 60]. Prussian blue is electrochemically reduced to form Prussian white (PW) which can then catalyze the reduction of H_2O_2 at low potentials of 0 V versus Ag/AgCl reference electrode [59].

6. Glutamate dehydrogenase–based biosensors

Another commonly used enzyme utilized in glutamate biosensors is L-glutamate dehydrogenase (GLDH) [49, 61–67]. GLDH (EC 1.4.1.2) catalyzes the deamination of amino acids, specifically the oxidative deamination of L-glutamate to 2-oxoglutarate (i.e., α-ketoglutarate) in the presence of nicotinamide adenine dinucleotide (NAD^+) which serves as an oxidized cofactor for the enzyme. In mammals, this reversible enzyme-catalyzed reaction strongly favors the formation of ammonium (NH_4^+) and 2-oxoglutarate.

$$\text{L-glutamate} + NAD^+ + H_2O \leftrightarrow \text{2-oxoglutarate} + NADH + NH_4^{\ +} + H^+ \qquad (2)$$

Glutamate biosensors have also been fabricated wherein both GLDH and GmOx are co-immobilized on the transducer surface [68]. Basu et al. co-immobilized both enzymes on a polycarbonate membrane by cross-linking procedures, involving glutaraldehyde, in the presence of a bovine serum albumin (BSA) spacer molecule in order to develop a biosensor for quantification of MSG in food. The MSG biosensor utilized substrate recycling which resulted in the amplification of the transducer response, thereby increasing the sensitivity [68].

On the other hand, rather than directly detecting the glutamate levels, Meng et al. quantified glutamate by the detection of the anodic current of enzymatically generated NADH [67]. The electron transfer kinetics of the oxidation of NADH is sluggish and the direct oxidation of NADH at bare electrodes requires a high overpotential (0.7–1.0 V), where many interferences can occur. Also, bare electrodes are more likely to be affected by fouling which is caused by the adsorption of oxidation products onto their surfaces. The authors overcame these challenges by preparing biocompatible biosensors utilizing thionine (Th) and single-walled carbon nanotubes (SWCNT) nanocomposite to catalyze the electrochemical oxidation of NADH at an anodic potential of less than 0.19 V versus a standard hydrogen electrode (SHE) [67].

Azmi et al. also developed a spectrophotometric biosensor based on GLDH that was immobilized in chitosan for the determination of ammonium in water samples [69]. Ammonium, which is known to be toxic even at low concentrations to various organisms, is widely used in the farming, chemical, and automotive industries. In addition, ammonium is used as a parameter in the assessment of drinking and industrial water quality. The authors immobilized GLDH in a chitosan film (a natural biopolymer which can found in the exoskeleton of crustaceans) and measured ammonium in water based on NADH oxidation in the presence of α-ketoglutaric acid. The biosensor had a detection limit of 0.005 mM and a linear range of 0.005–0.5 mM NH_4^+ [69].

Chitosan GLDH film is a popular enzyme immobilization matrix for biosensors due to certain advantageous properties of chitosan such as being nontoxic, biocompatible, biodegradable, an effective antibacterial, having high mechanical strength, good adhesion, and containing numerous amino and hydroxyl groups. Other glutamate sensing materials include polymer/enzyme composites [70, 71], nanoparticle iridium/carbon film [72], DNA-Cu(II)/polyamine membrane [73], nanoneedles [74], ferrocene functionalized SWCNT interdigitated construction film [52], Prussian blue film [45], and others.

7. Optimization of experimental conditions and measurement platforms

The topography of the prepared electrodes is usually studied using scanning electron microscope (SEM). SEM imaging allows for evaluation of uniformity and dispersity of the materials within the hybrid films. For example, it is possible to see if the composite materials are distributed uniformly over the entire surface of the electrode transducer. Agglomeration (i.e., sticking of particles to one another forming large groups) is sometimes observed in SEM images when the composites do not cover the surface uniformly. Porosity of the composite materials may also be assessed based on SEM images. The three-dimensional nanostructure morphology gives an indication whether the enzymes in the hybrid film will be accessible to the substrates. Modified electrodes are also much less likely to suffer from surface fouling, which results from adsorption of oxidation products to electrode surfaces when compared to bare electrodes.

As stated earlier, the incorporation of conductive nanomaterials, such as CNTs, into the hybrid film on a modified electrode surface is gaining popularity. CNTs often significantly improve electron transfer and kinetics [75]. This is due to the electrical properties of CNTs which vary significantly and depend on the structural differences between CNTs, resulting in some CNTs being highly conductive like metals, while others act more like semiconductors [76]. The current carrying capacity of certain CNTs can be up to 1000 times greater than that of a copper wire [77]. Furthermore, CNTs provide a tremendous increase in the surface area onto which enzymes or other biomolecules may be immobilized, which ultimately improves quantification of the chemical reaction that is being analyzed. CNTs are also ideal for use in biosensors as they are nontoxic, nanometer sized, strong, and chemically stable [76].

The electrochemical performance of the modified electrodes is often evaluated using redox reactions of benchmark species such as $Fe(CN)_6^{3-}/Fe(CN)_6^{4-}$ redox pair in cyclic voltammetry

(CV) studies. CV of this redox pair can be obtained using –300 mV initial potential, +800 mV switching potential, and –300 mV final potential with a carbon-based electrode against an Ag/AgCl reference electrode. The effect of changing scan rates (e.g., ranging from 30 to 120 mV/s) on electrochemical behavior is often also studied. Moreover, other neurotransmitters such as dopamine, which can directly undergo oxidation to dopamine-o-quinone at the electrode surface, may be detected in the living brain using voltammetry.

Characterization of modified electrode nanomaterials such as CNTs may be carried out by UV/Vis spectroscopy or Fourier transform infra-red (FTIR) spectroscopy. For example, Meng et al. utilized UV/Vis spectroscopy from 200 –800 nm [67] in their investigations. To confirm that thionine (Th) had adsorbed onto the SWCNT surface, the authors observed the occurrence of a notable absorption peak at about 600 nm as well as a small shoulder peak at circa 560 nm. To determine whether GLDH, the biocatalytic molecule, had been successfully immobilized on the Th/SWCNT nanocomposite, a peak at about 275 nm, which is an absorption peak characteristic for proteins, was monitored using UV spectroscopy.

When preparing a new biocatalytic sensor, parameters such as origin and availability of the enzyme, its operational and storage stability as well as immobilization procedure should also be carefully considered [78]. In addition, the influence of experiment conditions such as pH, temperature, ionic strength, and stirring on the enzyme-catalyzed reaction can be minimized by optimizing and keeping these conditions constant throughout the biosensor use. For example, GLDH-based biocatalytic sensors have utilized pHs ranging from 7.0 to 9.0 during the detection step [12, 67, 79]. Also, GmOx-based biosensors appear to utilize pHs from 7.0 to 7.4 [40, 68]. Of note, the optimal pH of the immobilized enzyme may be slightly different from the free enzyme in buffered aqueous solution. Furthermore, the detection temperatures which the enzymes are exposed to may vary depending on the specific application of the L-glutamate biosensor. The concentration of coenzyme NAD^+ will also need to be optimized for GLDH-based biosensors. Furthermore, the activity of enzymes such as GLDH (a hexameric enzyme from bovine liver) is affected by various cations and anions. Specifically, lanthanide (La^{3+}) and europium (Eu^{3+}) ions can enhance the activity of bovine GLDH, at least in solution [80]. Also, Meng et al. observed a decrease in catalytic currents generated by GLDH when Zn^{2+} was added to the solution with the inhibitory effect increasing with increasing concentrations of the zinc ions [67].

Amperometry is perhaps the most common electrochemical detection method used in quantitative analysis of an analyte once the biosensor has been characterized and optimized. It is a very popular electroanalytical detection method for quantitative analysis due to its simplicity and the low detection limits that can be achieved. In amperometry, the analyte concentration is determined by measurement of the signal—the current produced in a redox reaction as a function of time when a constant potential is applied to the electrodes. Amperometry results in current versus time plots where the current increases stepwise with each successive addition or formation of the redox active species. Amperometric signal response consisting of back-to-back steps in a "staircase" makes it relatively easy to identify the starting and final current for each analyte addition or the formation of redox active species. The response in amperometry is usually rapid and reaches a dynamic equilibrium which results in a steady-state current signal within seconds. The electron transfer of reactants during electrochemical oxidation is mainly determined by the conductivity of the working electrode material and the active functional groups on its surface. Three electrode systems with working, reference, and auxiliary electrodes are typically used in amperometry. In amperometry, the oxidation or reduction potential used for the detection step is characteristic of the analyte species, thus adding to selectivity of the method by eliminating interferences from other redox active species that may also be present in the sample. Also, the detection potential is stepped directly to the desired, optimum value, and current resulting from the redox reaction is detected by the transducer (the working electrode in the biosensor). Current generated by the reaction (i.e., the current passing through the electrochemical cell over time) is proportional to the concentration of the electroactive species in the sample. Sometimes, the charging current or background current (i.e., the current needed to apply the potential to the system) present at the beginning of each measurement requires some time in order to stabilize before quantitative measurements can be made using amperometry.

In order to be considered as an alternative to any existing and well-established instrumental analysis methods, such as high-performance liquid chromatography (HPLC) or spectrophotometry for quantification of L-glutamate in clinical or food applications, the performance of new biosensor devices has to be tested in serum or food samples, respectively. Ideally, no sample pretreatment other than dilution should be required. Also, the use of low-cost and disposable devices such as screen printed carbon electrodes (SPCE) is advantageous in the analysis of biological fluids where contamination may be a problem. Hughes et al. described the development and optimization of a disposable screen-printed amperometric biosensor for glutamate based on GLDH [79]. The authors also developed a stable, reagentless amperometric glutamate biosensor by incorporating the GLDH biorecognition components using a layer-by-layer deposition involving chitosan and MWCNTs on SPCE [81]. The new reagentless biosensor was applied to the measurement of glutamate in beef stock cubes and serum samples.

8. Figures of merit and the performance of glutamate biosensors

When new biosensors are being developed and characterized, it is common to report their detection limit, sensitivity, specificity, accuracy, reproducibility (i.e., precision), response times, reusability including recovery times, long-term stability (i.e., shelf-life), lifetime, and performance in real samples such as serum or soup [13, 82]. For electrochemical biosensors, reporting the electrode material and type as well as the transducer's surface area and detection potential versus a reference electrode are also important experimental details. **Table 1** summarizes many of these analytical figures of merit for selected glutamate biosensors allowing comparisons to be made between various enzyme-based biosensors.

Electrode	Configuration	Application	Enzyme	Optimal pH	Applied potential	Response time	LOD	Linear range	Ref.
Screen-printed graphite	CHIT/MB/SPCE	In vitro	GLDH	7	0.1 V vs. Ag/AgCl	2 s	1.5 μM	12.5–150 μM	Hughes [79]
Platinum microelectrode array	Silicon wafer/ polypyrrole/ Nafion	In vivo	GmOx	7.4	0.7 V vs. Ag/AgCl	<1 s	<1 μM	10–100 μM	Wassum [40]
Oxygen electrode	Glutaraldehyde/ bovine serum albumin	In vitro (food)	GmOx and GLDH	7		120 d	0.02 mg/L	0.02–1.2 mg/L	Basu [68]
Glassy carbon electrode	SWCNT/thionine	NADH	GLDH	8.3	0.19 V vs. Ag/ AgCl	5 s	0.1 μM	0.5–400 μM	Meng [67]
Au planar nanocomposite	CHIT/MWCNT/ ferricyanide	In vitro (food)	GLDH and diaphorase	9		<60 s	5.4 μM	10–3495 μM	Monošík [12]
SWCNT bundles	Ferrocene/SWCNT	In vitro			0.2 V vs. Ag/AgCl	<300 s	1 μM	1–7 μM	Huang [52]
Pt. microelectrode	CHIT/ceria & titania NPs	In vitro (hypoxic brain tissue)	GmOx		0.6 V vs. Ag/AgCl	2–5 s	0.594/0.493 μM		Özel [44]
Graphite	PB/graphite	Proof of concept	GmOx	6.5	−0.05 V vs. Ag/ AgCl	3 s	0.01 μM	0.01–0.1 mM	Liu [45]
Au electrode	cMWCNT/AuNP/ chitosan	In vitro (sera)	GmOx	7.5	0.135 V vs. Ag/ AgCl	2 s	1.6 μM	5–500 μM	Batra [83]
Vertically aligned CNT nanoelectrode array	VACNT-NEA	Proof of concept	GLDH		0 V vs. Ag/AgCl		57 nM	0.1–300 μM	Gholizadeh [75]
Pt electrode	Chitosan-glutaric dialdehyde gels	Proof of concept	GmOx		0.6 V vs. Ag/AgCl	2 s	0.10 μM	0.10–500 μM	Zhang [37]
Pt electrode	Chitosan	In vitro (food)	GmOx		0.4 V vs. Ag/AgCl	2 s	0.10 μM	1–10 μM	Zhang [38]
Patterned Pt thin film electrodes	Glutaraldehyde/ SiO_2/PCB	Cell culture fermentation	GmOx		0.6 V vs. Ag/AgCl		0.0002 μM	0.00022500 μM	Bäcker [41]

Electrode	Configuration	Application	Enzyme	Optimal pH	Applied potential	Response time	LOD	Linear range	Ref.
SAM on smart biodevice	ECD/thioglycolic acid	Proof of concept	GmOx		NA		0.089 μM	0.1–10,000 μM	Rahman [42]
Pt electrode	Glutaraldehyde	In vitro (brain tissue) uptake	GmOx	7.4	0.6 V vs. Ag/AgCl	15–20 s	0.5 μM	2–800 μM	Soldatkin [84]
Screen-printed graphite	MWCNT-CHIT-MB/CHIT-NAD$^+$-MB/MWCNT-CHIT-MB/MB-SPCE	In vitro proof of concept	GLDH	7	0.1 V vs. Ag/AgCl	<60 s	3 μM	7.5–105 μM	Hughes [81]

Table 1. A summary of previously published electrochemical glutamate biosensors, their electrode material, surface modification configuration, analytical figures of merit, and authors.

9. Conclusion

Enzyme-based electrochemical glutamate biosensors have tremendous potential for manufacturing of cost-efficient, easy-to-use, fast, and portable alternatives for a wide range of applications from medical/clinical testing or neurological studies for diagnostics involving this important neurotransmitter in vivo or in vitro to environmental monitoring, process monitoring, and food-sensing applications [13, 82]. Electrochemical detection schemes are also typically very simple, sensitive, independent of sample volume, and well suited for monitoring glutamate from nM to μM in real samples such as biological fluids or processed foods. The incorporation of the oxidase or dehydrogenase enzyme as the biorecognition component on the electrochemical transducer provides additional selectivity and in some detection schemes even significant signal enhancement. Many glutamate biosensors are label free while others incorporate various redox active mediator molecules into a composite material on the biosensor surface. An increasing number of enzyme-based glutamate biosensors also utilize the advantageous properties of nanomaterials such as biocompatible CNTs or metal nanoparticles. It is likely to take years or a decade before many of these biosensors described in this chapter go from proof-of-concept stage to mass production of inexpensive, small, and reliable devices capable of competing with existing instrument-intensive laboratory methods for glutamate quantification such as spectroscopy or chromatography. However, with continuous developments in molecular biology, nanofabrication methods, immobilization methods of biomolecules, and multiplexing capabilities, the production of sensitive, selective, fast, and easy-to-use biosensors for quantification of glutamate and other neurotransmitters will be feasible in the not too distant future.

Author details

Stanley L. Okon[1] and Niina J. Ronkainen[2*]

*Address all correspondence to: nronkainen@ben.edu

1 Presence Mercy Medical Center, Department of Psychiatry, Aurora, IL, USA

2 Department of Chemistry and Biochemistry, Benedictine University, Lisle, IL, USA

References

[1] Black DW, Andreasen NC. Introductory textbook of psychiatry (6th ed.). Arlington: American Psychiatric Publishing; 2014.

[2] Field JR, Walker AG, Conn PJ. Targeting glutamate synapses in schizophrenia.) Trends Mol Med. 2011;17:689–698.

[3] Paul IA, Skolnick P. Glutamate and depression: clinical and preclinical studies. Ann N Y Acad Sci. 2003;1003:250–272.

[4] Stahl, SM. Stahl's essential psychopharmacology (4th ed.). New York: Cambridge University Press; 2013.

[5] Javitt DC. Glutamate and schizophrenia: from theory to treatment implications. Psychiatr Times. 2014;31(9):49–52.

[6] Newsholme P, Procopio J, Lima MMR. Pithon-Curi TC, Curi, R. Glutamine and glutamate—their central role in cell metabolism and function. Cell Biochem Funct. 2003;21(1):1–9.

[7] Danbolt NC. Glutamate uptake. Progr Neurobiol. 2001;65(1):1–105.

[8] Featherstone DE. Intercellular glutamate signaling in the nervous system and beyond. ACS Chem Neurosci. 2009;1(1):4–12.

[9] Hu Y, Mitchell KM, Albahadily FN, Michaelis EK, Wilson GS. Direct measurement of glutamate release in the brain using a dual enzyme-based electrochemical sensor. Brain Res. 1994;659(1):117–125.

[10] Pasco N, Jeffries C, Davies Q, Downard AJ, Roddick-Lanzilotta AD, Gorton L. Characterization of a thermophilic L-glutamate dehydrogenase biosensor for amperometric determination of L-glutamate by flow injection analysis. Biosens Bioelectron. 1999;14:171–178.

[11] Jeffries C, Pasco N, Baronian K, Gorton L. Evaluation of a thermophilic enzyme for a carbon paste amperometric biosensor: L-glutamate dehydrogenase. Biosens Bioelectron. 1997;12(3):225–232.

[12] Monošík R, Streďanský M, Šturdík E. A biosensor utilizing L-glutamate dehydrogenase and diaphorase immobilized on nanocomposite electrode for determination of L-glutamate in food samples. Food Anal Method. 2013;6(2):521–527.

[13] Ronkainen NJ, Halsall HB, Heineman WR. Electrochemical biosensors. Chem Soc Rev. 2010;39(5):1747–1763.

[14] Valero E, Garcia-Carmona F. A continuous spectrophotometric method based on enzymatic cycling for determining L-glutamate. Anal Biochem. 1998;259(2):265–271.

[15] Khampha W, Meevootisom V, Wiyakrutta S. Spectrophotometric enzymatic cycling method using L-glutamate dehydrogenase and D-phenylglycine aminotransferase for determination of L-glutamate in foods. Anal Chim Acta. 2004;520(1):133–139.

[16] Blankenstein G, Preuschoff F, Spohn U, Mohr KH, Kula MR. Determination of L-glutamate and L-glutamine by flow-injection analysis and chemiluminescence detection: comparison of an enzyme column and enzyme membrate sensor. Anal Chim Acta. 1993;271(2):231–237.

[17] Dawson LA, Stow JM, Palmer AM. Improved method for the measurement of glutamate and aspartate using capillary electrophoresis with laser induced fluorescence detection and its application to brain microdialysis. J Chromatogr B. 1997;694(2):455–460.

[18] Tivesten A, Lundqvist A, Folestad S. Selective chiral determination of aspartic and glutamic acid in biological samples by capillary electrophoresis. Chromatographia, 1997;44 (11–12):623–633.

[19] Lada MW, Vickroy TW, Kennedy RT. Evidence for neuronal origin and metabotropic receptor-mediated regulation of extracellular glutamate and aspartate in rat striatum in vivo following electrical stimulation of the prefrontal cortex. J Neurochem. 1998;70(2):617–625.

[20] Nakanishi H. Improved cleanup and derivatization for gas chromatographic determination of monosodium glutamate in foods. J Assoc Off Anal Chem. 1983;66(6):1528–1531.

[21] Buck K, Voehringer P, Ferger B. Rapid analysis of GABA and glutamate in microdialysis samples using high performance liquid chromatography and tandem mass spectrometry. J Neurosci Methods. 2009;182(1):78–84.

[22] Piepponen TP, Skujins A. Rapid and sensitive step gradient assays of glutamate, glycine, taurine and γ-aminobutyric acid by high-performance liquid chromatography–fluorescence detection with o-phthalaldehyde–mercaptoethanol derivatization with an emphasis on microdialysis samples. J Chromatogr B. 2001;757(2):277–283.

[23] Afzal A, Afzal M, Jones A, Armstrong D (2002). Rapid determination of glutamate using HPLC technology. In Oxidative stress biomarkers and antioxidant protocols (pp. 111–115). Humana Press. New York City, NY.

[24] Yao T, Okano G. Simultaneous determination of L-glutamate, acetylcholine and dopamine in rat brain by a flow-injection biosensor system with microdialysis sampling. Anal Sci. 2008;24(11):1469–1473.

[25] Hascup KN, Hascup ER, Pomerleau F, Huettl P, Gerhardt GA. Second-by-second measures of L-glutamate in the prefrontal cortex and striatum of freely moving mice. J Pharmacol Exp Ther. 2008;324(2):725–731.

[26] Rutherford EC, Pomerleau F, Huettl P, Strömberg I, Gerhardt GA. Chronic second-by-second measures of l-glutamate in the central nervous system of freely moving rats. J Neurochem. 2007;102(3):712–722.

[27] Villarta RL, Cunningham DD, Guilbault GG. Amperometric enzyme electrode for the determination of L-glutamate. Talanta. 1991;38:49–55.

[28] Ghobadi S, Csöregi E, Marko-Varga G, Gorton L. Bienzyme carbon paste electrodes for L-glutamate determination. Curr Sep. 1996;14:94–102.

[29] Belay A, Ruzgas T, Csöregi E, Moges G, Tessema M, Solomon T, Gorton L. LC-biosensor system for the determination of the neurotoxin β-N-oxalyl-L-α,β-diaminopropionic acid. Anal Chem. 1997;69:3471–3475.

[30] Yao T, Nanjyo Y, Nishino H. Micro-flow in vivo analysis of L-glutamate with an on-line enzyme amplifier based on substrate recycling. Anal Sci. 2001;17:703–708.

[31] Moser I, Jobst G, Urban GA. Biosensor arrays for simultaneous measurement of glucose, lactate, glutamate, and glutamine. Biosens Bioelecton. 2002;17(4):297–302.

[32] Mikeladze E, Collins A, Sukhacheva M, Netrusov A, Csöregi E. Characterization of a glutamate biosensor based on a novel glutamate oxidase integrated into a redox hydrogel. Electroanalysis. 2002;14(15–16):1052–1059.

[33] O'Neill RD, Chang SC, Lowry JP, McNeil CJ. Comparisons of platinum, gold, palladium and glassy carbon as electrode materials in the design of biosensors for glutamate. Biosens Bioelecton. 2004;19:1521–1528.

[34] Qhobosheane M, Wu D, Gu Y, Tan W. A two-dimensional imaging biosensor to monitor enhanced brain glutamate release stimulated by nicotine. J Neurosci Method. 2004;135(1):71–78.

[35] Castillo J, Isik S, Blochl A, Rodrigues NP, Bedioui F, Csoregi E, Schuhmann W, Oni J. Simultaneous detection of the release of glutamate and nitric oxide from adherently growing cells using an array of glutamate and nitric oxide selective electrode Biosens Bioelectron. 2005;20:1559–1565.

[36] Castillo J, Blochl A, Dennison S, Schuhmann W, Csoregi E. Glutamate detection from nerve cells using a planer electrodes array integrated in a microtiter plate. Biosens Bioelectron. 2005;20:2116–2119.

[37] Zhang M, Mullens C, Gorski W. Chitosan-glutamate oxidase gels: synthesis, characterization, and glutamate determination. Electroanalysis. 2005;17(23):2114–2120.

[38] Zhang M, Mullens C, Gorski W. Amperometric glutamate biosensor based on chitosan enzyme film. Electrochim Acta. 2006;51(21):4528–4532.

[39] Schuvailo OM, Soldatkin OO, Lefebvre A, Cespuglio R, Soldatkin AP. Highly selective microbiosensors for in vivo measurement of glucose, lactate and glutamate. Anal Chim Acta, 2006;573:110–116.

[40] Wassum KM, Tolosa VM, Wang J, Walker E, Monbouquette HG, Maidment NT. Silicon wafer-based platinum microelectrode array biosensor for near real-time measurement of glutamate in vivo. Sensors. 2008;8:5023–5036.

[41] Bäcker M, Delle L, Poghossian A, Biselli M, Zang W, Wagner P, Schöning MJ. Electrochemical sensor array for bioprocess monitoring. Electrochim Acta. 2011;56(26):9673–9678.

[42] Rahman MM. Fabrication of mediator-free glutamate sensors based on glutamate oxidase using smart micro-devices. J Biomed Nanotechnol. 2011;7(3):351–357.

[43] Batra B, Kumari S, Pundir CS. Construction of glutamate biosensor based on covalent immobilization of glutamate oxidase on polypyrrole nanoparticles/polyaniline modified gold electrode. Enzyme Microb Technol. 2014;57:69–77.

[44] Özel RE, Ispas C, Ganesana M, Leiter JC, Andreescu S. Glutamate oxidase biosensor based on mixed ceria and titania nanoparticles for the detection of glutamate in hypoxic environments. Biosens Bioelectron. 2014;52: 397–402.

[45] Liu L, Shi L, Chu Z, Peng J, Jin W. Prussian blue nanocubes modified graphite electrodes for the electrochemical detection of various analytes with high performance. Sens Actuators B. 2014;202:820–826.

[46] Kusakabe H, Midorikawa Y, Fujishima T, Kuninaka A, Yoshino H. Purification and properties of a new enzyme, L-glutamate oxidase from *Streptomyces* sp. X-119-6 grown on wheat bran. Agric Biol Chem 1983;47(6):1323–1328.

[47] Böhmer A, Müller A, Passarge M, Liebs P, Honeck H, Müller HG. A novel L-glutamate oxidase from *Streptomyces endus*. Eur J Biochem. 1989;182(2):327–332.

[48] Ryan M, Lowry J, O'Neill R. Biosensor for neurotransmitter L-glutamic acid designed for efficient use of L-glutamate oxidase and effective rejection of interference. Analyst. 1997;122(11):1419–1424.

[49] Alvarez-Crespo SL, Lobo-Castañón MJ, Miranda-Ordieres AJ, Tuñón-Blanco P. Amperometric glutamate biosensor based on poly(o-phenylenediamine) film electrogenerated onto modified carbon paste electrodes. Biosens Bioelecton. 1997;12:739–747.

[50] Garguilo MG, Huynh N, Proctor A, Michael AC. Amperometric sensors for peroxide, choline, and acetylcholine based on electron transfer between horseradish peroxidase and a redox polymer. Anal Chem. 1993;65(5):523–528.

[51] Burmeister JJ, Gerhardt GA. Self-referencing ceramic-based multisite microelectrodes for the detection and elimination of interferences from the measurement of L-glutamate and other analytes. Anal Chem. 2001;73(5):1037–1042.

[52] Huang XJ, Im HS, Lee DH, Kim HS, Choi YK. Ferrocene functionalized single-walled carbon nanotube bundles. Hybrid interdigitated construction film for L-glutamate detection. J Phys Chem C. 2007;111(3):1200–1206.

[53] Qiu JD, Deng MQ, Liang RP, Xiong M. Ferrocene-modified multiwalled carbon nanotubes as building block for construction of reagentless enzyme-based biosensors. Sens Actuat B. 2008;135(1):181–187.

[54] Claussen JC, Franklin AD, Fisher TS, Porterfield DM. U.S. Patent No. 8,715,981. Washington, DC: U.S. Patent and Trademark Office; 2014.

[55] Oldenziel WH, Dijkstra G, Cremers TI, Westerink BH. Evaluation of hydrogel-coated glutamate microsensors. Anal Chem. 2006;78(10):3366–3378.

[56] Karyakin AA, Karyakina EE, Gorton L. Amperometric biosensor for glutamate using Prussian blue-based "artificial peroxidase" as a transducer for hydrogen peroxide. Anal Chem. 2000;72(7):1720–1723.

[57] Sitnikova NA, Komkova MA, Khomyakova IV, Karyakina EE, Karyakin AA. Transition metal hexacyanoferrates in electrocatalysis of H_2O_2 reduction: an exclusive property of Prussian Blue. Anal Chem. 2014;86(9):4131–4134.

[58] Dale N, Tian F. U.S. Patent No. 8,417,314. Washington, DC: U.S. Patent and Trademark Office; 2013.

[59] Cinti S, Arduini F, Moscone D, Palleschi G, Killard AJ. Development of a hydrogen peroxide sensor based on screen-printed electrodes modified with inkjet-printed Prussian blue nanoparticles. Sensors. 2014;14(8):14222–14234.

[60] Gong H, Sun M, Fan R, Qian L. One-step preparation of a composite consisting of graphene oxide, Prussian blue and chitosan for electrochemical sensing of hydrogen peroxide. Microchim Acta. 2013;180(3–4):295–301.

[61] Amine A, Kauffmann JM. Preparation and characterization of a fragile enzyme immobilized carbon paste electrode. Bioelectrochem Bioenerg. 1992;28:117–125.

[62] Amine A, Kauffmann JM, Palleschi G. Investigation of the batch injection analysis technique with amperometric biocatalytic electrodes using a modified small-volume cell. Anal Chim Acta. 1993;273:213–218.

[63] Cosford RJO, Kuhr WG. Capillary biosensor for glutamate. Anal Chem. 1996;68:2164–2169.

[64] Chakraborty S, Retna RC. Amperometric biosensing of glutamate using carbon nanotube based electrode. Electrochem Commun. 2007;9(6):1323–1330.

[65] Tang L, Zhu Y, Xu L, Yang X, Li C. Amperometric glutamate biosensor based on self-assembling glutamate dehydrogenase and dendrimer-encapsulated platinum nanoparticles onto carbon nanotubes. Talanta. 2007;73(3), 438–443.

[66] Tang L, Zhu Y, Yang X, Li C. An enhanced biosensor for glutamate based on self-assembled carbon nanotubes and dendrimer-encapsulated platinum nanobiocomposites-doped polypyrrole film. Anal Chim Acta. 2007;597(1):145–150.

[67] Meng L, Wu P, Chen G, Cai C, Sun Y, Yuan Z. Low potential detection of glutamate based on the electrocatalytic oxidation of NADH at thionine/single-walled carbon nanotubes composite modified electrode. Biosens Bioelectron. 2009;24(6):1751–1756.

[68] Basu AK, Chattopadhyay P, Roychudhuri U, Chakraborty R. A biosensor based on co-immobilized L-glutamate oxidase and L-glutamate dehydrogenase for analysis of monosodium glutamate in food. Biosens Bioelectron. 2006;21(10):1968–1972.

[69] Azmi NE, Ahmad M, Abdullah J, Sidek H, Heng LY, Karuppiah N. Biosensor based on glutamate dehydrogenase immobilized in chitosan for the determination of ammonium in water samples. Anal Biochem. 2009;388(1):28–32.

[70] Rahman MA, Kwon NH, Won MS, Choe ES, Shim YB. Functionalized conducting polymer as an enzyme-immobilizing substrate: an amperometric glutamate microbiosensor for in vivo measurements. Anal Chem. 2005;77(15):4854–4860.

[71] McMahon CP, Rocchitta G, Serra PA, Kirwan SM, Lowry JP, O'Neill RD. Control of the oxygen dependence of an implantable polymer/enzyme composite biosensor for glutamate. Anal Chem. 2006;78(7):2352–2359.

[72] You T, Niwa O, Kurita R, Iwasaki Y, Hayashi K, Suzuki K, Hirono S. Reductive H_2O_2 detection at nanoparticle iridium/carbon film electrode and its spplication as L-glutamate enzyme sensor. Electroanalysis. 2004;16(1–2):54–59.

[73] Hasebe Y, Gu T, Kusakabe H. Glutamate biosensor using a DNA-Cu (II)/polyamine membrane as a novel electrocataytic layer for cathodic determination of hydrogen peroxide. Electrochemistry (Tokyo). 2006;74(2):179.

[74] Boo H, Jeong RA, Park S, Kim KS, An KH, Lee YH, Han JH, Kim HC, Chung TD. Electrochemical nanoneedle biosensor based on multiwall carbon nanotube. Anal Chem. 2006;78(2):617–620.

[75] Gholizadeh A, Shahrokhian S, Iraji zad A, Mohajerzadeh S, Vosoughi M, Darbari S, Koohsorkhi J, Mehran M. Fabrication of sensitive glutamate biosensor based on vertically aligned CNT nanoelectrode array and investigating the effect of CNTs density on the electrode performance. Anal Chem. 2012;84(14):5932–5938.

[76] Ronkainen NJ, Okon SL. Nanomaterial-based electrochemical immunosensors for clinically significant biomarkers. Materials. 2014;7(6):4669–4709.

[77] Fulekar MH. Nanotechnology: importance and applications. New Delhi: I.K. International Pvt Ltd.; 2010.

[78] Bartlett PN, editor. Bioelectrochemistry fundamentals, experimental techniques and applications. West Sussex: John Wiley & Sons; 2008.

[79] Hughes G, Pemberton RM, Fielden PR, Hart JP. Development of a disposable screen printed amperometric biosensor based on glutamate dehydrogenase, for the determination of glutamate in clinical and food applications. Anal Bioanal Electrochem. 2014;6(4):435–449.

[80] Zhuang QK, Dai HC, Gao XX, Xin WK. Electrochemical studies of the effect of lanthanide ions on the activity of glutamate dehydrogenase. Bioelectrochemistry. 2000;52(1):37–41.

[81] Hughes G, Pemberton RM, Fielden PR, Hart JP. Development of a novel reagentless, screen-printed amperometric biosensor based on glutamate dehydrogenase and NAD^+, integrated with multi-walled carbon nanotubes for the determination of glutamate in food and clinical applications. Sens Actuators B. 2015;216, 614–621.

[82] Bahadır, E. B., & Sezgintürk, M. K. (2015). Applications of commercial biosensors in clinical, food, environmental, and biothreat/biowarfare analyses. *Analytical Biochemistry*, 478, 107–120.

[83] Batra, B., & Pundir, C. S. (2013). An amperometric glutamate biosensor based on immobilization of glutamate oxidase onto carboxylated multiwalled carbon nanotubes/gold nanoparticles/chitosan composite film modified Au electrode. *Biosensors and Bioelectronics*, 47, 496–501.

[84] Soldatkin, O., A. Nazarova, N. Krisanova, A. Borysov, D. Kucherenko, I. Kucherenko, N. Pozdnyakova, A. Soldatkin, and T. Borisova. (2015) "Monitoring of the velocity of high-affinity glutamate uptake by isolated brain nerve terminals using amperometric glutamate biosensor." *Talanta* 135, 67–74.

Chapter 3

Graphene-Paper Based Electrochemical Sensors

Minwei Zhang, Arnab Halder, Xianyi Cao,
Chengyi Hou and Qijin Chi

Additional information is available at the end of the chapter

http://dx.doi.org/10.5772/intechopen.68186

Abstract

Graphene paper as a new form of graphene-supported nanomaterials has received worldwide attention since its first report in 2007. Due to their high flexibility, lightweight and good electrical conductivity, graphene papers have demonstrated the promising potential for crucial applications in electrochemical sensors and energy technologies among others. In this chapter, we present some examples to overview recent advances in the research and development of two-dimensional (2D) graphene papers as new materials for electrochemical sensors. The chapter covers the design, fabrication, functionalization and application evaluations of graphene papers. We *first* summarize the mainstream methods for fabrication of graphene papers/membranes, with the focus on chemical vapour deposition techniques and solution-processing assembly approaches. A large portion of this chapter is *then* devoted to the highlights of specific functionalization of graphene papers with polymer and nanoscale functional building blocks for electrochemical-sensing purposes. In terms of electrochemical-sensing applications, the emphasis is on enzyme-graphene and nanoparticle-graphene paper-based systems for the detection of glucose. We *finally* conclude this chapter with brief remarks and outlook.

Keywords: graphene paper, 2D-layered nanomaterials, electrochemical sensor, metal oxide, polymer, nanoparticle

1. Introduction

Precise monitoring of chemical or biological processes is of extreme importance for medical and biological applications. Electrochemical sensors can ideally fulfil that goal by converting a chemical or a biological response into a processable and quantifiable signal. In the past two decades, intensive research and development of electrochemical sensors have enabled to fabricate different types of devices. After the development of many successful commercial electrochemical sensors in the classic configurations, currently there is a notable transition and increasing demands for the development of flexible and wearable sensors. The development of flexible electrochemical sensors depends crucially on the discovery and preparation of free-standing and flexible new materials.

Flexible electronic devices, especially flexible electrochemical sensors, have become an advanced technology with the aim at solving some tremendous real-world challenges, such as in situ sensing for health and environmental problems. Compared with conventional devices, flexible electrode-based devices are versatile and highly adaptive. For example, such devices occupy less space and could be compatible with any shape-targeted systems such as human body and rough or irregular-shaped substrates. Ideal flexible electrodes should provide high conductivity for sensing electrons and thermal energy, high stability and excellent mechanical strength. In this regard, carbon nanotube (CNT)-based devices had dominated research and development attention before [1, 2], because of the unique one-dimensional nano-channel mechanical strength. However, one has come to realize that CNT-based flexible electrodes are largely limited to fundamental interests because of several serious technical challenges, such as high cost of producing CNTs, complex process of obtaining high-density vertically aligned CNTs, chemical and biocompatibility concerns and difficulties in achieving large-scale production, which can hardly overcome. On the other hand, the recent rise of synthesis and processing of two-dimensional (2D) graphene and its derivatives offer a new and promising opportunity for developing a novel class of flexible electrodes most likely with required physiochemical properties. Graphene paper, as one of the derivatives from graphene or graphene oxide (GO)-supported building blocks, was introduced by Ruoff and co-workers in 2007 [3]. Due to their remarkable mechanical property, tunable conductivity and versatile functionality, graphene papers could promote the research and development of new-generation flexible electrode-based sensors and other electronic devices. This chapter aims at highlighting some recent examples using graphene papers as electrode materials for developing flexible and ultrasensitive electrochemical sensors.

2. Synthesis of graphene paper-based electrodes

Thanks to the intrinsic planar structure of graphene sheets, it is feasible to effectively control the periodic alignment of graphene nanosheets into 2D graphene papers/films/membranes through ionic interaction, hydrophobic effect, hydrogen bonding and/or π-π stacking [4, 5]. The resulting 2D graphene materials not only retain some of the key properties of individual graphene sheets but also create new or collective properties resulting from their unique structure. Moreover, a wide range of functionalized materials, such as metal, polymer, metal oxide (MO) and other semiconductors, can be loaded into 2D-layered GO papers, endowing the graphene papers with multiple functionalities required by specific applications [4–8]. Given the excellent physicochemical properties of graphene paper and tunable functional groups, these 2D architectures have been widely studied for their applications in water purification, biomimetics, sensors, flexible electrodes, energy conversion and storage and optoelectronic devices [7–11]. As outlined below, several crucial methods have been developed to assemble the layered structure of graphene papers.

2.1. Chemical vapour deposition

Chemical vapour deposition (CVD) is a chemical process used to produce high quality, high performance and solid materials. The CVD technique is a direct and promising method for the preparation of large-scale, near-perfect graphene films on various substrates. The graphene films obtained by CVD methods displayed high flexibility, transparency and electrical conductivity [12, 13], which is attractive for various electronic devices. Commonly, transition metals, such as Ge [14, 15], Ni [16, 17], Cu [18, 19], Rh [20–22] and Co [23, 24], are used as the substrates for the CVD growth of graphene films. Due to very low carbon solubility in Cu, ease of etching and feasibility of high-quality graphene transfer, Cu has emerged as a favoured substrate chosen mostly for the growth of large-area graphene films [25], which was first introduced by Ruoff's research group in 2009 [13]. Large-area and single-layer graphene films with a low percentage (less than 5%) of the areas having few layers can be deposited on copper substrates by CVD process using methane as a carbon source (**Figure 1a**). The graphene film was robust enough to be transferred onto Si/SiO_2 substrates for fabrication of dual-gated field-effect transistors. They also demonstrated that as-synthesized graphene films could be transferred to other substrates (**Figure 1b** and **c**). To meet some special requirements, non-metal materials, such as Si [14, 15], SiO_2 [26–28], BN [29, 30] and Si_3N_4 [31, 32], were also used as substrates. However, the non-metal substrates showed the drawback limitations including slow growth rate and discontinuous size. For the CVD growth of graphene films on a metal substrate, it is necessary to mention that a transfer step for the following integration of graphene film into any solid-state electronic devices is usually required. Given the polycrystalline nature of the film and roughness of metal foils, industrial roll-to-roll transfer technology could be employed. Moreover, free-standing and element-doped (e.g. N and S) [33] graphene films can also be prepared directly by the CVD technique. However, the conventional etching method could be satisfied at laboratory scales, but it is very challenging for the large-scale preparation of high-quality graphene films arising from high-handling skills and time-consuming setback.

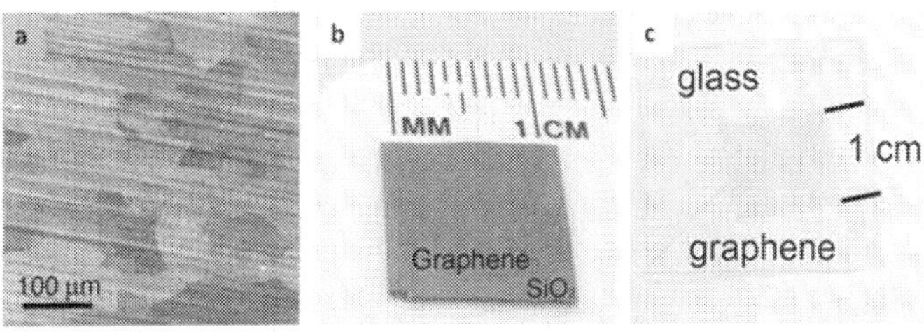

Figure 1. (**a**) SEM image of a graphene film on a copper foil after 30 min CVD growth. (**b** and **c**) Digital photographs of the CVD graphene film transferred onto a SiO_2/Si substrate and a glass plate, respectively [13].

2.2. Solution-processed assembly

Owing to oxygen containing polar functional groups existing on the surface and at the edges, GO is well dispersible in many polar solvents including water [34]. This structural feature and chemical capability facilitate GO as a favourite starting material for the assembly of various graphene architectures such as graphene papers and three-dimensional (3D) graphene sponges.

2.2.1. *Vacuum filtration*

To build a well-organized 2D macroscopic structure using single-layer GO sheets as a start material, Ruoff's group introduced a facile method for the fabrication of graphene paper. That is, GO sheets could be assembled into a paper-like material under a directional flow. Vacuum filtration of colloidal dispersions of GO sheets through an anodic membrane filter yielded free-standing GO papers with tunable thickness and mechanical property in a dried state. The fracture edges of a GO paper sample were imaged by scanning electron microscopy (SEM) to reveal its well-ordered layers over almost the entire cross section of the paper sample. The sandwiched structure between less densely packed 'wavy' skin layers was approximately 100–200 nm thick. In a typical GO paper specimen, the layer-to-layer distance (d-spacing) was measured as approximately 0.83 nm by the X-ray diffraction (XRD) spectrum [3]. As a result of its facile processability, vacuum filtration has been popularly used to prepare free-standing graphene papers. For example, cellular graphene paper was prepared by the following three steps [35]. *Firstly,* direct filtration of partially reduced GO with ascorbic acid mixture solution was performed. *Then,* the vacuum was disconnected immediately once no free dispersion was left on the filter paper and both the filter membrane and partially reduced GO (rGO) paper were vertically immersed into a liquid nitrogen bath for a rapid freezing about 30 min. *Finally,* graphene paper was further reduced by thermal reduction to enhance its electrical conductivity. From low-magnification SEM image of the as-prepared porous graphene paper (**Figure 2a**), the paper has a thickness of about 12.6 μm with continuous open networks. Thanks to ice-crystal templating, graphene paper also displayed honeycomb-like structures with the pore sizes varying in the range of hundreds of nanometres to several micrometres, as shown by the high-magnification SEM images (**Figure 2c** and **d**). The porous walls consisting of thin layers of graphene sheets were also clearly imaged by both SEM and transmission electron microscopy (TEM) (**Figure 2a**, **e** and **f**). High-resolution TEM (HRTEM) images further revealed that there are many crumpled 5–10 nm graphene sheets stacked on the surface of the graphene walls that are several tens of nanometres thick [35]. Our group has used GO as a starting material and further functionalized it with *Prussian Blue* (PB) nanoparticles (PBNPs) [36] or PB-shelled Au nanoparticles, Au@PBNPs [37] to develop a new kind of free-standing and flexible graphene papers. These PB-functionalized rGO papers are highly flexible and electroactive, and they were tested for use as disposable non-enzymatic electrochemical sensors.

2.2.2. *Layer-by-layer assembly*

Self-assembly is a popular approach towards cost-effective preparation of thin films, which has also been applied for fabricating multilayer graphene papers with tunable composition

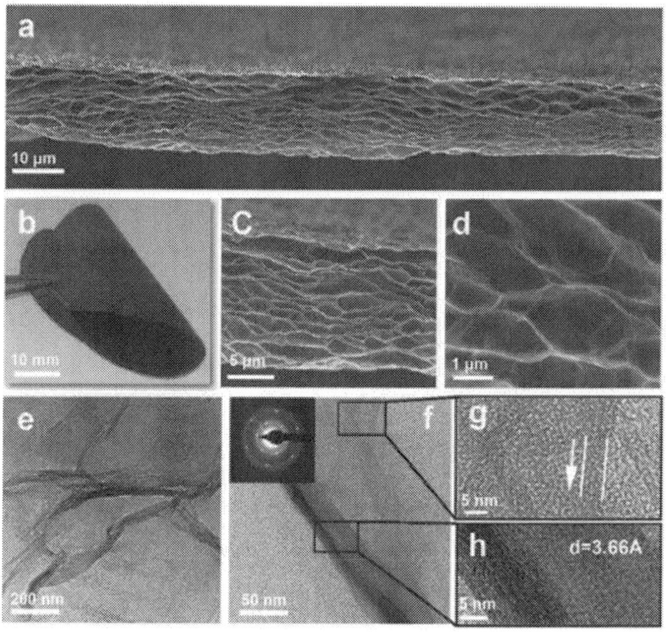

Figure 2. Structural features of rGO paper: (**a**) Cross-sectional SEM image of rGO paper after complete reduction. (**b**) Digital photograph of as-prepared rGO paper. (**c** and **d**) High-resolution cross-sectional SEM images of rGO paper. (**e**) TEM image of a part of graphene paper. (**f–h**) HRTEM images of the porous walls in the graphene paper. The inset image in (**f**) is the selected area diffraction pattern of the porous wall [35].

and architecture [8, 10, 38]. The driving forces involved in layer-by-layer (LBL) assembly generally include hydrogen bonding, electrostatic interaction and covalent bonding [8, 39, 40]. Zou and Kim reported a diffusion-driven LBL assembly process to prepare graphene-based architecture and demonstrated its application for the construction of GO sheets into various 3D macrostructures [41]. This assembly process is driven by the complexation of the negatively charged GO sheets and positively charged branched polyethylenimine (b-PEI) at a defined interface. The key step for assembling GO sheets to GO paper is that the diffusion of b-PEI molecules allows the complex to continuously grow into foam-like frameworks with tunable porosity. In a typical experimental process as shown in **Figure 3**, a small amount of the b-PEI solution was *first* dropped on certain substrates, for example, a glass plate (**Figure 3a**) or a filter paper (**Figure 3e**). *Then*, the b-PEI containing substrate was immersed in a GO suspension and left on a shaker for 24 h. After the formation of films, the samples were purified by dialysis in distilled (DI) water for 2–3 days. In the *final* step, the GO film was dried by normal heating or by freeze-drying. To their surprise, the GO film obtained by heating dry is a dense, tightly packed multilayer film (**Figure 3d**), which can be explained by the surface tension of water pulling the GO sheets together during evaporation. By contrast, freeze-drying the sample minimized such effects and helped preserve the porous structure (**Figure 3c**). Furthermore, the pore size could be adjusted by using different solvents [41].

3. Structural and morphologic characterization

As-synthesized graphene papers are a very promising candidate for energy device and sensor applications. Structural characterization is of paramount importance for understanding the correlation between their nanostructures and performances. Therefore, some advanced techniques are intensively used to characterize graphene paper-based materials, such as atomic force microscopy (AFM), Raman spectroscopy, Fourier transform infrared spectroscopy (FTIR), X-ray photoelectron spectroscopy (XPS), energy dispersive X-ray spectroscopy (EDX) and thermogravimetric analysis (TGA). In particular, electronic microscopies (SEM and TEM) are among the most powerful techniques to reveal the structural details of nanostructured graphene papers. For example, a combination of TEM and SEM was used to systematically characterize Au@PBNPs hybrid graphene papers. It is shown that Au@PB NPs were well distributed on graphene sheet surface (**Figure 5**), and cross-sectional SEM images indicated that the Au@PB NPs were successfully doped into the interlayers of the graphene paper to form a layer-by-layer sandwiched structure (**Figure 6**) [37].

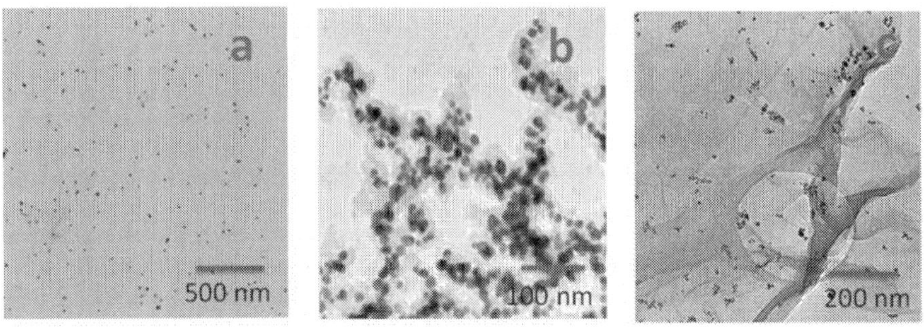

Figure 5. TEM images of different types of nanoparticles in the free form and in the immobilized form on GO sheets: (**a**) AuNPs; (**b**) core/shell Au@PBNPs and (**c**) core/shell Au@PBNPs on GO sheets [37].

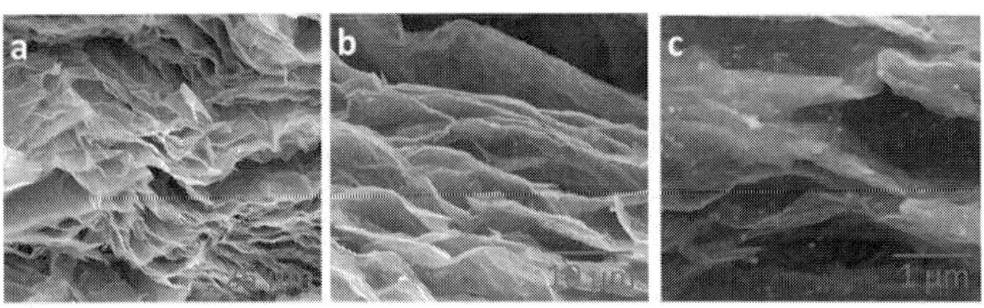

Figure 6. Cross-sectional SEM images of Au@PBNP-functionalized rGO papers with various magnifications [37].

4. Functionalization of graphene papers

The preparation and applications of graphene paper-based materials are attracting huge interests due to their unique electronic, optical, magnetic, thermal and mechanical properties. Compared to pure graphene papers, functionalized graphene papers or membranes can offer notable advantages ascribed to their hybrid structure, specific functionality and improved physicochemical properties. In general, graphene papers could be functionalized with a variety of chemical components such as metallic materials, metal oxides, polymers and supramolecular units, via various methods. Here, we only focus on three types of functional derivation, that is, metallic nanostructures, metal compounds and polymers, which are in favour of electrochemical-sensing applications.

4.1. Metallic nanostructures

Nanostructured metals are attracting intensive interests from the scientific community, owing to their fabulous properties and diversity of applications [57, 58]. A number of studies have explored the cooperation and synergistic effects between nanostructured metals and graphene paper, such as Pt nanoparticles (PtNPs) [59], Pt nanowires (PtNWs) [60], Au@Pt core-shell NPs [61], Au nanoparticles (AuNPs) [62] and Ag nanoparticles (NPs) [63]. As we mentioned in Section 2.1, CVD has made a great contribution to the preparation of graphene films. Hydrocarbon gases, through CVD consolidating on metal surfaces, enable the formation of a uniform and large-size graphene film, which can then be transferred onto solid substrates decorated with metallic NPs [64]. For example, Du and co-workers fabricated an AuNP-modified graphene film by a CVD-thermal release tape method, which has been used as an active substrate for the surface-enhanced Raman scattering (SERS) detection of analytes [65]. Furthermore, to enhance the electrical properties and surface plasmon signal of the graphene film, Ag nanowire was deposited further on the graphene film by a two-step procedure. In general, graphene films were *first* prepared by a CVD method, and then Ag nanowire was deposited on the graphene film by either physical loading [66], physical vapour deposition (PVD) [67], spin coating [68] or electrochemical pulse deposition [63].

Self-assembly provides a simple and time-saving way to prepare large-scale nanostructure-modified graphene paper by precisely controlling the experimental parameters. For example, AgNP-graphene hybrid paper was assembled through electrostatic self-assembly. *Firstly*, polycation-modified AgNPs were synthesized and *then* conjugated to a GO sheet. The reduction of GO and absorption of AgNPs occurred simultaneously, resulting in the formation of AgNP-graphene hybrid paper [69]. With the similar strategy, a homogeneous mixture of a GO suspension with AuNPs allowed self-assembly at the air/liquid interface to form a multi-layered GO-AuNP composite film [70]. As shown in **Figure 7a**, UV-vis absorption spectra of GO, AuNP, GO-AuNP and GO-SWCNT solutions indicate that AuNPs did not influence the π-electronic transitions in the aromatic carbon-carbon of GO, but the electrostatic repulsion between the deprotonated carboxylic groups at the edge of GO nanosheet and negatively charged AuNPs modulated the electronic environment of AuNPs, causing the blue shift of the absorption of

AuNPs by 4 nm. The resulting rGO-AuNP film (**Figure 7b**) showed an enhanced electrode kinetics and a cyclic voltammetric response in proportion to the amount of AuNPs incorporated. An enhancement of anodic peak current was clearly observed compared with that of the pure rGO films [70]. Zhang et al. developed a one-step strategy to synthesize self-assembled AuNP-graphene hybrid paper [71]. A mixture solution containing both GO and $HAuCl_4$ was directly used as a starting material. The two precursors were reduced into rGO and AuNPs, respectively, by glucose under heat treatment, and rGO and AuNPs *then* self-assembled into a multi-layered paper structure upon solvent evaporation. *Finally*, a PET film was used to harvest the formed AuNP-graphene hybrid paper [71]. Moreover, rGO/$Ni(OH)_2$ paper was prepared by the EPD process, which offered a facile, rapid, scalable and environmentally friendly method for making graphene paper [72]. $NiCo_2O_4$ was grown on the surfaces of porous N-doped graphene sheets through hydrogen bonding, van der Waals forces or covalent bonding with the functional groups of graphene such as –COOH groups. The as-prepared hierarchically porous graphene paper with $NiCo_2O_4$ displayed a remarkable catalytic activity towards oxygen evolution reaction (OER) [73].

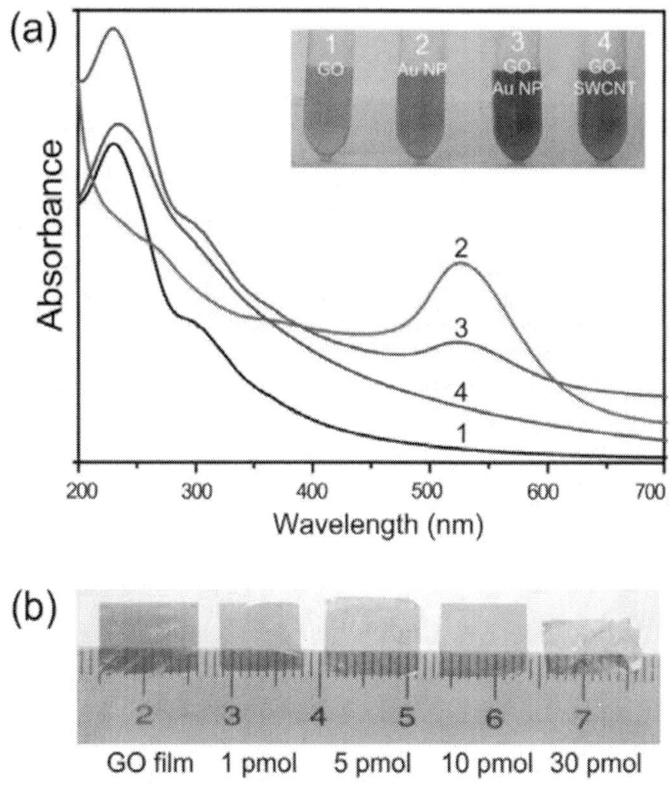

Figure 7. (**a**) UV-vis absorption spectra and digital images (inset) of 1: GO, 2: AuNP, 3: GO-AuNP and 4: GO–SWCNT solution. (**b**) Top-layered GO-AuNP composite films with a variation of the AuNP concentration from 1 to 30 p mol [70].

N-doped graphene-SnO_2 papers were prepared by inducing the 7,7,8,8-tetracyanoquinodimethane anion as both the nitrogen source and a complexing agent to formulate a sandwich structure, which exhibited a large capacity, high-rate capability and excellent cycling stability for electrochemical energy storage, such as lithium-ion batteries [74]. As noted, metallic nanostructure-modified graphene papers could also be prepared by the EPD process or vacuum filtration method [8, 75].

4.2. Metallic compound nanostructures

Metallic compounds, especially transition metal oxides (TMOs) or transition metal hydroxides (TMHs), are a key family of materials in a variety of current demanding needs for sensor, catalysis, energy storage and conversion, optical electronics and piezoelectric mechanics, attributed to their nature of versatile functionality [76]. Given such a fact, metallic compound-functionalized graphene papers have been a hot topic in recent years. Commonly, metal oxide and GO were synthesized separately, and then MO-graphene paper was fabricated through mixing the two (or more) components and applying a typical vacuum filtration process. For example, our group has reported the fabrication of free-standing graphene–*Prussian* blue (PB) composite paper through the filtration of chemically compatible graphene–PB nanohybrids, with the overall procedure shown in **Figure 8**. PB nanoparticles were *first* attached into rGO by the electrostatic attraction, and the hybrid material was *then* directly assembled into 2D flexible graphene paper sensor [36]. SnO_2 nanosphere hybrid graphene paper was prepared by direct vacuum filtration method. Using the similar strategy, Fe_2O_3 NPs [77], $Na_{2/3}Fe_{1/2}Mn_{1/2}O_2$ NPs [78] and MnO_2 NPs [79] were successfully combined with graphene paper. In addition, graphene/MnO_2 paper could be fabricated by filtration of the mixture of GO with $Mn(NO_3)_2$ and $KMnO_4$ solution and was then subjected to the reduction of GO to rGO [80]. In some cases, nanostructured metallic compounds only functionalized the graphene-paper surface rather than the interlayers. Such metallic compound-functionalized graphene hybrid materials were obtained mainly by the growth of nanostructured metallic compounds on the surface of graphene films by various techniques. For example, pulse laser deposition was applied to deposit V_2O_5 film on the graphene paper surface to yield flexible energy storage devices [81]. NiO [82], TiO_2 [83–85], indium-gallium-zinc oxide [86] and PtNPs/MnO_2 nanowires [61] were all successfully deposited onto graphene paper surfaces for specific functionalization.

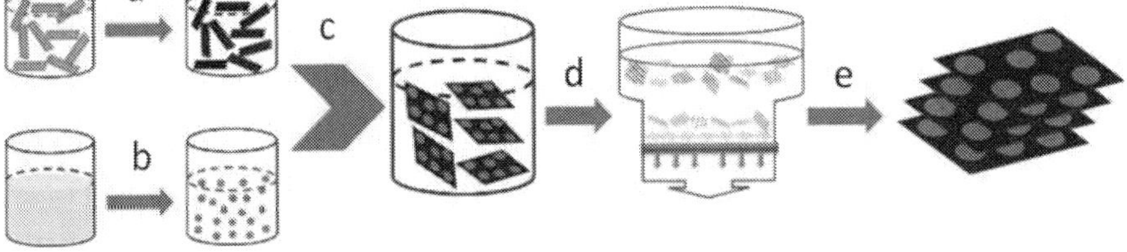

Figure 8. Schematic illustration of the preparation procedure for PBNP-functionalized graphene papers: (**a**) chemical reduction of GO to rGO, (b) synthesis procedure of PNBPs, (**c**) preparation of PBNPs–rGO hybrid nanosheets, (**d**) vacuum filtration to produce graphene paper and (e) annealing process to enhance the paper conductivity [36].

In addition, non-metallic nanostructures such as those carbon-containing building blocks (e.g. CNTs, fullerenes, carbon black, carbon fibre and carbon spheres), Si, SiO_2, Si_3N_4 and others [7, 10, 51] were also widely used to functionalize graphene papers.

4.3. Polymers

Besides inorganic nanocomposites, polymer-functionalized graphene papers also have widely been used for sensor and energy conversion and storage applications. In general, the polymer could be stabilized with graphene paper through the two approaches: (1) covalent bonding. Because GO contains a notable number of oxygen-containing groups, it provides remarkable opportunities for further modification with polymer either by 'grafting to' or by 'grafting from' techniques. (2) Non-covalent attachment such as electrostatic attraction, hydrogen bonding or π–π stacking; a variety of polymers have been incorporated into graphene-based films with a uniform multilayer structure and novel functionality, such as polyaniline [87–91], poly(sodium 4-styrenesulfonate) [92] and poly(p-phenylene vinylene) [48]. Recently, Mu et al. developed a series of PVDF-modified graphene monolayer papers with a gradient rGO/GO structure. In the gradient graphene paper, the GO region could respond to the environment stimuli including changes in humidity, temperature or light, leading to shrinking or swelling of the GO sheets [93]. In another example, poly-dopamine was used to functionalize graphene paper, the as-prepared graphene paper displayed high performance of responses to moisture, heat and light. The response of this water-driven

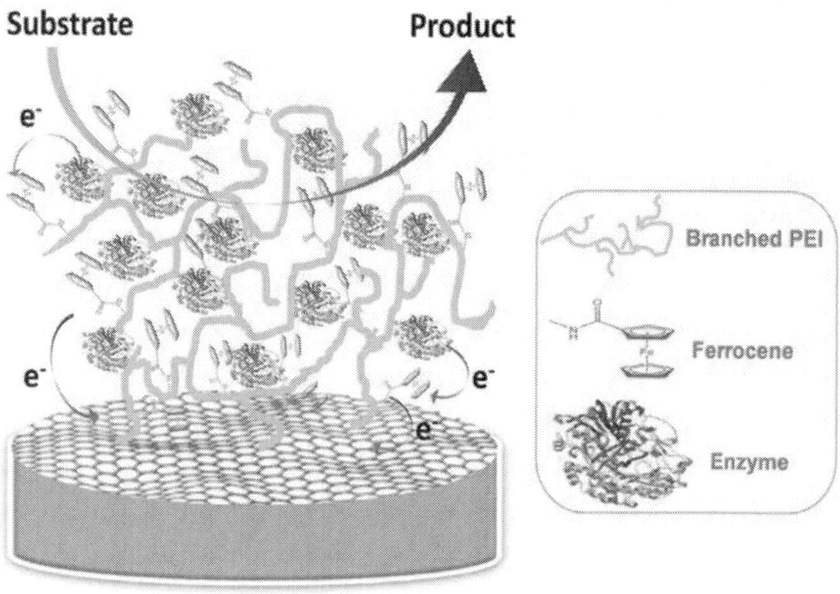

Figure 9. Schematic illustration of structure, electron transfer and bioelectrocatalysis of Fc-rGOP-enzyme hybrid systems [95].

actuator to multiple stimuli allows the artificial muscles and electric generators to be fabricated [94]. In our recent work, as shown in **Figure 9**, a highly branched PEI (b-PEI) was used for reduction and simultaneous derivation of GO to form a biocompatible polymeric matrix on rGO nanosheet. Ferrocene redox moieties were then grafted onto rGO nanosheets through the polymer matrix. The as-prepared functional composite is electrochemically active and enabled to accommodate enzymes stably. For proof-of-the concept studies, two crucial redox enzymes for biosensors (i.e. cholesterol oxidase and glucose oxidase (GOx)) were used to test the platform with good outcomes [95].

5. Applications of graphene paper as electrochemical-sensing platforms

Graphene papers with the advantages of low price, high quality and simple synthesis process have a potential to apply in electronic and optoelectronic devices, electrochemical energy devices, water treatment and sensors [7, 8, 10, 50, 51]. Especially the self-assembled 2D graphene papers/films/membranes functionalized with nanostructured metals, polymers and biomolecules are promising candidates for sensing applications, as summarized in **Table 1**, which is an overview of enzyme-based and non-enzymatic graphene papers/films sensors for the detection of glucose. In particular, in the field of electrochemical biosensors with high sensitivity, these candidates have found widespread uses in clinic diagnosis, environment monitoring and for

Composite material	Electrocatalyst	Linear range	LOD	Sensitivity ($\mu A\ cm^{-2}\ mM^{-1}$)	References
PBNPs/GP/GOD	GOD	6-8 mM	10 μM	25	[36]
MnO_2/PtNPs/GP	PtNPs	10-46000 μM	20 μM	52.36	[60]
AuNPs/GP	AuNPs	100-30000 μM	5 μ M	58.54	[61]
AuNPs/PANI/GP/GOD	GOD	200-11200 μM	100 μM		[88]
Fc/GP/GOD	GOD	100-15500 μM	5 μM	3.45	[95]
CuO NC/GP	CuO	2-4000 μM	0.7 μM	1360	[100]
CuO/S-GP	CuO	100-10500 μM	0.08 μM	1298	[101]
CS/N-GP/GOD	GOD	200-1800 μM	64 μM	10.5	[111]
AuNPs/GP/GOD	GOD	20-2260 μM	4.1 μM	3.855	[112]
ZrO_2/GP/GOD	GOD	200-1600 μM	45.6 μM	7.6	[113]
AuNPs/PBNPs/CS/GP/GOD	GOD	25-3200 μM	10 μM	58.7	[114]
GOD/CS/Fc/GO	GOD	20-6780 μM	7.6 μM	10	[115]
TEOS/APTES/Fc/CS/GP/GOD	GOD	20-5390 μM	6.5 μM	19.5	[116]
NiNPs/GP	NiNPs	1-110 μM	-	813	[117]
NiO/GP	NiO	20-2100 μM	0.1 μM	1020	[118]
PtNPs/GP	PtNPs	2-2100 μM	1 μM	1508	[119]
CuO/GP	CuO	1-8000 μM	1 μM	1065	[120]
Co_3O_4/gP	Co_3O_4	up to 80 μM	25 nM	669.78	[121]

PANi, polyaniline; GP, graphene paper or graphene film; GOD, glucose oxidase; CS, chitosan; Fc, ferrocene.

Table 1. Overview of enzyme-contained and non-enzymatic graphene paper/film sensors for detection of glucose.

quality control in industrial, food and agricultural products [9, 34, 59, 61, 96]. In Section 5, we mainly address the applications of graphene papers as different types of electrochemical sensor.

5.1. Enzyme-graphene paper sensors

Graphene paper displayed outstanding physical and electrochemical properties, such as high conductivity, large surface area, abundant defect sites, superior electrocatalytic activity and fast electron transfer rate, hence various enzyme biosensors based on this attractive material have been fabricated recently [11, 95]. We have successfully fabricated graphene paper doped with chemically compatible PBNPs as a nanohybrid electrocatalyst [36]. *Prussian* blue was loaded on the graphene nanosheet by electrostatic attraction. The as-prepared PBNPs–rGO papers are further capable of biocompatible accommodation of enzymes for the development of free-standing enzyme-based biosensors.

The correlation between the electrocatalytic current and glucose concentration shows a linear relation up to about 6–8 mM with a sensitivity of 25 $\mu A\ mM^{-1}\ cm^{-2}$ and the detection limit down to about 10 μM. Furthermore, our group had recently explored a facile way for the successful synthesis of redox active and bioengineering of rGO for the development of versatile biosensing platform [95]. We developed a simple way for the synthesis of ferrocene (Fc) functionalization of highly b-PEI-linked rGO as well as the development of biocompatible matrix for accommodation of different bio-recognition elements. Ferrocene redox moieties were attached to rGO nanosheets through PEI. The polymer acts as both a reducing agent and a molecular spacer for rGO. Fc-PEI-rGO is electrochemically active and can offer a biocompatible microenvironment for immobilization of cholesterol oxidase and glucose oxidase. The as-constructed electroactive matrix was further used for the development of integrated biosensing platforms for cholesterol and glucose sensing. As measured, the current-time (I-t) and calibration curves are shown in **Figure 10**, the detection limit of the biosensors for glucose and cholesterol is 5 and 0.5 μM ($S/N = 3$), respectively. The linear response range of the biosensor is from 0.1 to 15.5 mM for glucose. Furthermore, this biosensing platform shows good anti-interference ability and reasonable stability. The nanohybrid biosensing materials can be further combined with screen-printed electrodes, which were successfully used for measuring the glucose level of real human serum samples. In addition to these two enzymes, other enzymes can also be immobilized onto graphene-based electrodes for the construction of various enzyme biosensors, including horseradish peroxidase (HRP), alcohol dehydrogenase (ADH), catalase and urease [87, 97–99].

5.2. Non-enzymatic sensors for detection of small biomolecules

Glucose oxidase-modified electrode is the most common class of amperometric biosensors for glucose detection because GOx enables catalytic oxidation of glucose with high sensitivity and selectivity. However, enzyme-modified electrodes have some disadvantages such as the instability of the electrode and unsatisfactory reproducibility, complicated enzyme immobilization procedure and high cost of enzymes. For addressing such challenging issues, non-enzymatic glucose biosensors based on graphene papers have been explored. As mentioned above, nanostructured MO displayed high performance for the fabrication of electrochemical sensors. The working principles of using MO as an electrochemical catalyst for the detection of glucose can be described by Eqs. (1) and (2). A couple of redox peaks are obtained in a basic solution, and MO(OH) can then oxidize glucose to glucolactone, leading to a remarkable increase in the anodic peak current. Graphene paper plays a key role as supporting electrode and speeding up electron transfer reaction. Several studies have reported that MO-functionalized graphene papers can sensitively detect glucose without any biological materials involved in the fabrication of sensors. For example, CuO, a p-type semiconductor with a narrow band gap of 1.2 eV, is very promising for the development of glucose sensors because of high specific surface area, good electrochemical activity and the possibility of promoting electron transfer reactions at a lower overpotential. For example, CuO nanocubes were deposited on a graphene film by the EPD process, under the optimized conditions, a linear

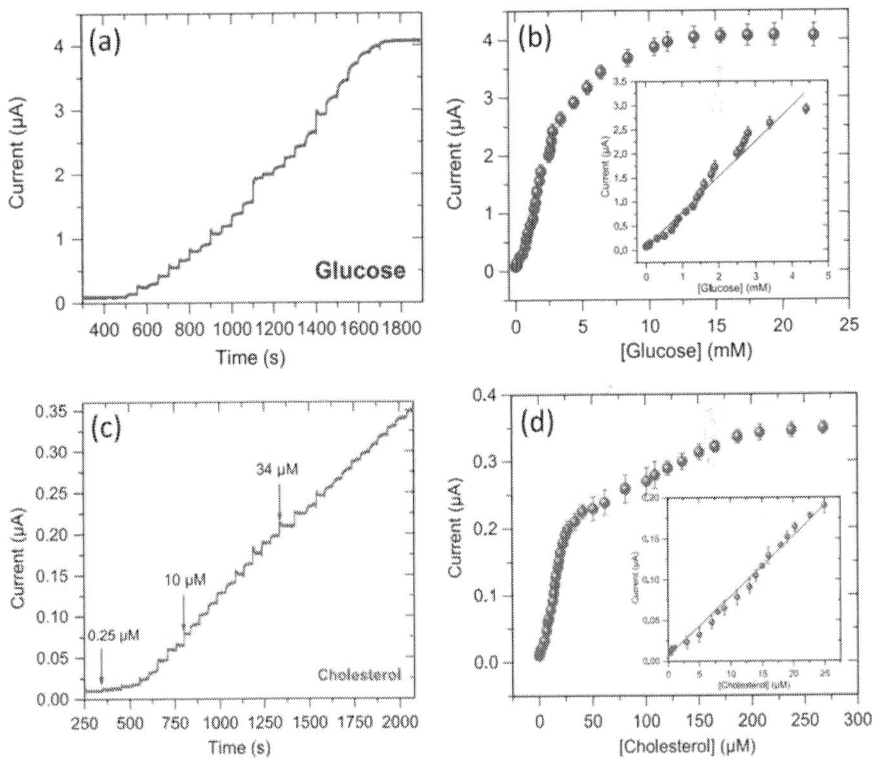

Figure 10. (**a**) Current-time (*I-t*) curve obtained at Fc-rGOP-GOx/GCE electrode upon successive injection of glucose, (**b**) dependence of electrocatalytic currents on the glucose concentration, (**c**) amperometric responses to cholesterol with Fc-rGOP-ChOx/GCE electrode and (**d**) calibration curve for cholesterol. The insets in (**b**) and (**d**) are the linear part of current responses to the substrates. Electrolyte: 10 mM PBS (pH 7.0); the working electrode potential fixed at 0.4 V (vs SCE); substrate (either glucose or cholesterol) solution was injected at a regular time interval to the supporting electrolyte which was under stirring [95].

range up to 4 mM with a sensitivity of 1360 μA mM^{-1} cm^{-2} at a positive potential (i.e. 0.55 V) was achieved [100]. In order to improve the performance of CuO-modified graphene paper as non-enzymatic sensor, CuO-functionalized S-doped graphene was successfully synthesized through a facile microwave-assisted approach. The CuO/S-doped graphene-based sensor exhibited a rapid response of 2s (**Figure 11**), a wide linear range of 0.1–10.5 mM, a high sensitivity of 1298.6 μA mM^{-1} cm^{-2} and a low detection limit of 80 nM. Moreover, this biosensing platform showed good anti-interference ability and reasonable stability (**Figure 11**) [101]. In another case, PtNPs/MnO_2 nanowires/graphene paper used as a free-standing paper electrode for non-enzymatic detection of glucose, the sensor response is linear to the glucose concentrations in the range from 0.1 to 30.0 mM with a detection limit of 0.02 mM (S/N = 3) and detection sensitivity of 58.54 μA cm^{-2} mM^{-1} [61]. Moreover, nanostructured NiO [102] and CoO [103] were loaded on graphene paper through various advanced techniques for the fabrication of ultrasensitive non-enzymatic sensors

$$MO + OH^{-} \rightarrow MO(OH) + e- \quad (1)$$

$$MO(OH) + glucose \rightarrow MO + gluconolactone \quad (2)$$

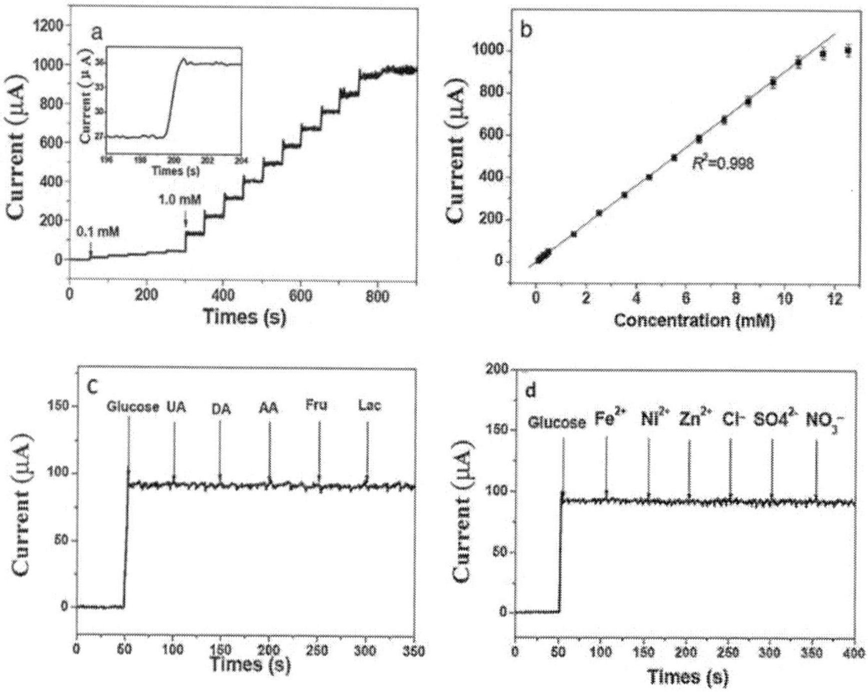

Figure 11. (**a**) Amperometric responses to the successive addition of glucose in 0.1 M NaOH at 0.50 V (the inset showing the response time upon the addition of 0.1 mM glucose). (**b**) The calibration curve of the current response to glucose. (**c**) Interference test with continuous injections of glucose (1.0 mM) and UA, DA, AA, fructose and lactose (with the same concentration of 0.1 mM). (**d**) Interference tests with continuous injections of 1.0 mM glucose and 0.10 mM of other ions [101].

http://dx.doi.org/10.5772/intechopen.68186

Ascorbic acid (AA), dopamine (DA), uric acid (UA) and other small biomolecules are among the key small biomolecules that affect human health. Therefore, electrochemical sensing of those small biomolecules has played a vital role in monitoring the health. Recently, building electrochemical-sensing platforms based on graphene-supported nanomaterials for the detection of those small biomolecules has been a hot topic, due to the fast electron transfer kinetics and superior electrocatalytic activity of graphene-based materials. Dopamine is one of the mostly studied small biomolecules using graphene-based biosensors. Determination of dopamine, ascorbic acid and uric acid concentration is crucial, because the concentration level of these molecules in human body is closely linked to the health status. Graphene film-based electrodes as electrochemical sensors have shown high performance towards the analysis of dopamine, ascorbic acid and uric acid [104–110]. For example, Xia and co-workers developed a multifunctional electrochemical sensor based on N-doped graphene (NG), which can be used to simultaneously determine AA, DA and UA [105].

N-doped graphene exhibited high performance for electrochemical sensing of AA, DA and UA due to its unique structure and physiochemical property. The electrochemical sensor had a wide linear response to AA, DA and UA [105]. Moreover, some important small molecules for food safety control also could be detected by using graphene paper-based electrodes. As an example, an Orange II sensor was fabricated based on PtNP-functionalized rGO film. This sensor was evaluated for the detection of Orange II in an acetate buffer. The results showed a wide linear range (1×10^{-8}–6×10^{-7} M) and the detection limit could be down to 3.4×10^{-10} M (at $S/N = 3$). The as-prepared electrochemical sensor also showed high selectivity, impressive stability and promising responses to real samples with an excellent recovery [59].

5.3. Electrochemiluminescence sensors

Electrochemiluminescence (ECL) is a kind of luminescence induced by electrochemical reactions normally at liquid/solid interfaces. ECL has proven to be a highly sensitive and selective method and is a very useful tool in analytical science. The ECL method combines the advantages of chemiluminescent analysis with the ease of reaction control by applying an electrode potential. Therefore, ECL-based sensors have recently been developed using graphene papers. For example, based on $Ru(bpy)_3^{2+}$ and alcohol dehydrogenase-modified graphene/bovine serum albumin composite film, ethanol biosensor was developed [122]. In another case, Li et al. developed an ultrasensitive ECL sensor for the detection of DA. In this study, cationic polythiophene derivative, poly[3-(1,10-dimethyl-4-piperidinemethylene)thiophene-2,5-diyl chloride] (PTh-D), was used to improve stability, and NH_2–graphene acted as electron transfer accelerator. Under the optimal experimental conditions, the ECL signal linearly decreased with the increase of DA concentration in the range of 0.1–50.0 mM with a detection limit of 0.04 mM (**Figure 12**). This simple prepared ECL sensor also exhibited high selectivity, good reproducibility and long-term stability. The proposed ECL sensor was also evaluated by detecting DA in real samples [123]. Xu et al. obtained a paper-based solid-state ECL sensor based on poly(sodium 4-styrenesulfonate)-functionalized graphene/Nafion composite film. Very impressively, the detection limit could be down to ($S/N = 3$) 5.0 nM. The sensor exhibited excellent reproducibility and long-term stability. Moreover, highly porous Fe_3O_4 nanocrystal clusters were added to poly(sodium 4-styrenesulphonate) hybrid graphene paper for further

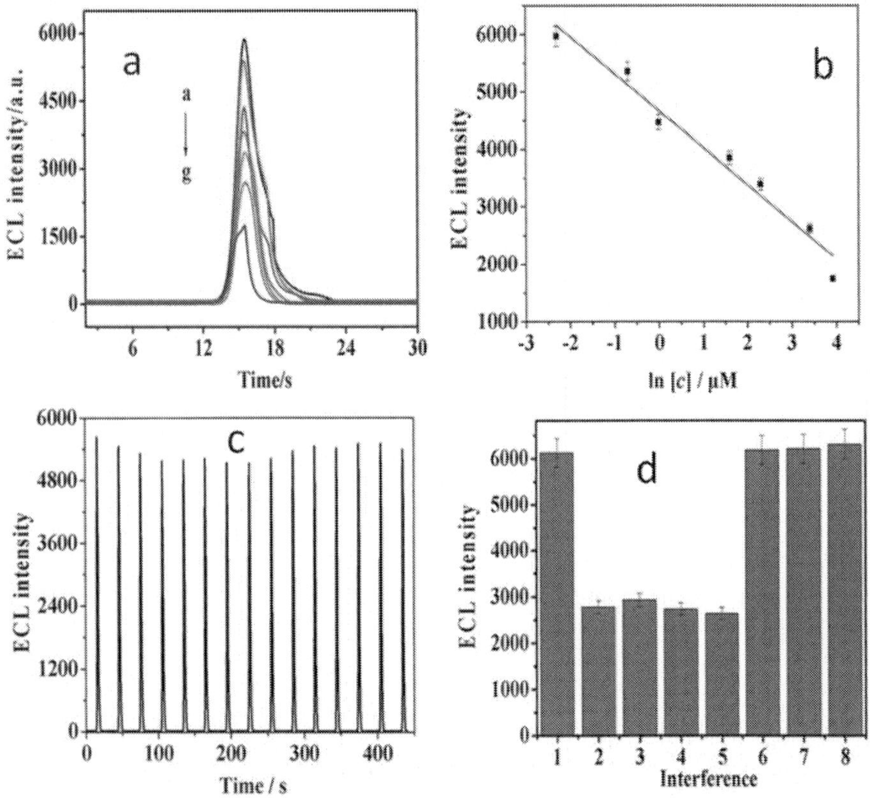

Figure 12. (**a**) The response of ECL to DA, from (a) to (g): 0.1 to 50 mM. (**b**) Corresponding calibration curve of response ECL versus DA concentrations. (**c**) The ECL intensity of the sensor in a solution containing 0.1 M KCl and 0.5 mM DA under continuous scanning for 15 cycles. (**d**) The interference tests of the ECL sensor [123].

enhancing the specific ability of graphene paper. This type of graphene paper-based sensors showed a high performance to detect those specific compounds containing tertiary amino groups and DNA with guanine and adenine [92, 124].

5.4. Others

Immunosensors are among the most important sensors for biomedical applications. Due to high flexibility and conductivity, the development of graphene paper-based electrochemical immunosensors is highly desirable [125]. Indeed, there are some striking examples in this regard. For example, a simple and label-free electrochemical impedimetric immunosensor for immunoglobulin G (IgG) based on chemically modified graphene was prepared by the Martin group [126]. In another example, an efficient, low cost and robust impedimetric immunosensor for the rapid and sensitive detection of *Escherichia coli* O157:H7 (*E. coli* O157:H7) was developed by using AuNP-functionalized free-standing graphene paper electrodes.

Electrochemical impedance spectroscopy (EIS) was used to detect *E. coli* O157:H7 captured on the paper electrode, and a broad linear range (1.5×10^2–1.5×10^7 cfu mL^{-1}) and a low detection limit (1.5×10^2 cfu mL^{-1}) were achieved [127].

6. Concluding remarks and outlook

In summary, worldwide researchers have explored graphene paper-based sensors by exploiting their unique advantages including high sensitivity, conductivity and in situ sensing. The recent research advances suggest that graphene paper-based materials could play a significant role in developing flexible sensors and electronic devices due to their intriguing structural and functional features. However, the progresses are accompanied by new challenges. Among the possibly critical challenges, the biocompatibility, electronic conductivity and stretchable capability are likely mostly concerned. In this regard, the further development of graphene papers by incorporation of biocompatible and conducting polymers is particularly encouraged. Although there are these challenges to be overcome in order to meet practically capable applications in real world, we would like to believe that flexible and free-standing graphene papers should continue emerging as a widely useful material platform for developing electrochemical-sensing technologies in the near future.

Acknowledgements

This work was supported by DFF_FTP, the Danish Research Council for Technology and Product Science (to Q.C., Project No. 12-127447). M.Z. acknowledges the CSC PhD scholarship (No. 201306170047).

Author details

Minwei Zhang, Arnab Halder, Xianyi Cao, Chengyi Hou and Qijin Chi*

*Address all correspondence to: cq@kemi.dtu.dk

Department of Chemistry, Technical University of Denmark, Kongens Lyngby, Denmark

References

[1] Wang J, Musameh M. Carbon nanotube/Teflon composite electrochemical sensors and biosensors. Analytical Chemistry. 2003;**75**:2075–2079. doi:10.1021/ac030007+

[2] Ahammad AJS, Lee J-J, Rahman MA. Electrochemical sensors based on carbon nanotubes. Sensors. 2009;**9**:2289–2319. doi:10.3390/s90402289

[3] Dikin DA, Stankovich S, Zimney EJ, Piner RD, Dommett GHB, Evmenenko G, Nguyen ST, Ruoff RS. Preparation and characterization of graphene oxide paper. Nature 2007; **448**:457-60. doi:10.1038/nature06016

[4] Wu D, Zhang F, Liang H, Feng X. Nanocomposites and macroscopic materials: Assembly of chemically modified graphene sheets. Chemical Society Review. 2012;**41**:6160. doi:10.1039/c2cs35179j

[5] Xu Y, Shi G. Assembly of chemically modified graphene: Methods and applications. Journal of Material Chemistry. 2011;**21**:3311. doi:10.1039/c0jm02319a

[6] Joshi RK, Alwarappan S, Yoshimura M, Sahajwalla V, Nishina Y. Graphene oxide: The new membrane material. Applied Material Today. 2015;**1**:1–12. doi:10.1016/j.apmt.2015.06.002

[7] Liu G, Jin W, Xu N. Graphene-based membranes. Chemical Society Review. 2015;**44**:5016–5030. doi: 10.1039/C4CS00423J

[8] Cong H-P, Chen J-F, Yu S-H. Graphene-based macroscopic assemblies and architectures: An emerging material system. Chemical Society Review. 2014;**43**:7295–325. doi:10.1039/c4cs00181h

[9] Gao H, Duan H. 2D and 3D graphene materials: Preparation and bioelectrochemical applications. Biosensors and Bioelectronics. 2015;**65**:404–419. doi:10.1016/j.bios.2014.10.067

[10] Shao Y, El-Kady MF, Wang LJ, Zhang Q, Li Y, Wang H, Mousavi MF, Kaner RB. Graphene-based materials for flexible supercapacitors. Chemical Society Review. 2015;**44**:3639–3665. doi:10.1039/C4CS00316K

[11] Zhang Y, Shen J, Li H, Wang L, Cao D, Feng X, Liu Y, Ma Y, Wang L. Recent progress on graphene-based electrochemical biosensors. The Chemical Record. 2016;**16**:273–294. doi:10.1002/tcr.201500236

[12] Lee DH, Kim JE, Han TH, Hwang WJ, Jeon SW, Choi SY, Hong SH, Lee WJ, Ruoff RS, Kim SO. Versatile carbon hybrid films composed of vertical carbon nanotubes crown on mechanically compliant graphene films. Advanced Materials. 2010;**22**:1247–1252. doi:10.1002/adma.200903063

[13] Li X, Cai W, An J, Kim S, Nah J, Yang D, Piner R, Velamakanni A, Jung I, Tutuc E, Banerjee SK, Colombo L, Ruoff RS. Large area synthesis of high quality and uniform graphene films on copper foils. Science. 2009;**324**:1312–1314. doi:10.1126/science.1171245

[14] Pasternak I, Wesolowski M, Jozwik I, Lukosius M, Lupina G, Dabrowski P, Baranowski JM, Strupinski W. Graphene growth on Ge(100)/Si(100) substrates by CVD method. Scientific Reports. 2016;**6**:21773. doi:10.1038/srep21773

[15] Dabrowski J, Lippert G, Avila J, Baringhaus J, Colambo I, Dedkov YS, Herziger F, Lupina G, Maultzsch J, Schaffus T, Schroeder T, Kot M, Tegenkamp C, Vignaud D, Asensio M-C. Understanding the growth mechanism of graphene on Ge/Si(001) surfaces. Scientific Reports. 2016;**6**:31639. doi:10.1038/srep31639

[16] Baraton L, He ZB, Lee CS, Cojocaru CS, Châtelet M, Maurice J-L, Lee YH, Pribat D. On the mechanisms of precipitation of graphene on nickel thin films. EPL. 2011;**96**:46003. doi:10.1209/0295-5075/96/46003

[17] Losurdo M, Giangregorio MM, Capezzuto P, Bruno G. Graphene CVD growth on copper and nickel: Role of hydrogen in kinetics and structure. Physical Chemistry Chemical Physics. 2011;**13**:20836. doi:10.1039/c1cp22347j

[18] Zhang J, Wang Z, Niu T, Wang S, Li Z, Chen W. Elementary process for CVD graphene on Cu(110): Size-selective carbon clusters. Scientific Reports. 2014;**4**:4431. doi:10.1038/srep04431

[19] Niu T, Zhou M, Zhang J, Feng Y, Chen W. Growth intermediates for CVD graphene on Cu(111): Carbon clusters and defective graphene. Journal of the American Chemical Society. 2013;**135**:8409–8414. doi:10.1021/ja403583s

[20] Gotterbarm K, Zhao W, Höfert O, Gleichweit C, Papp C, Steinrück H-P. Growth and oxidation of graphene on Rh(111). Physical Chemistry Chemical Physics. 2013;**15**:19625. doi:10.1039/c3cp53802h

[21] Liu M, Gao Y, Zhang Y, Zhang Y, Ma D, Ji Q, Gao T, Chen Y, Liu Z. Single and polycrystalline graphene on Rh(111) following different growth mechanisms. Small. 2013;**9**:1360–1366. doi:10.1002/smll.201202962

[22] Kordatos A, Kelaidis N, Giamini SA, Marquez-Velasco J, Xenogiannopoulou E, Tsipas P, Kordas G, Dimoulas A. AB stacked few layer graphene growth by chemical vapor deposition on single crystal Rh(1 1 1) and electronic structure characterization. Applied Surface Science. 2016;**369**:251–256. doi:10.1016/j.apsusc.2016.02.023

[23] Ago H, Ito Y, Mizuta N, Yoshida K, Hu B, Orofeo CM, Tsuji M, Ikeda KI, Mizuno S. Epitaxial chemical vapor deposition growth of single-layer graphene over cobalt film crystallized on sapphire. ACS Nano. 2010;**4**:7407–7414. doi:10.1021/nn102519b

[24] Cabrero-Vilatela A, Weatherup RS, Braeuninger-Weimer P, Caneva S, Hofmann S. Towards a general growth model for graphene CVD on transition metal catalysts. Nanoscale. 2016;**8**:2149–2158. doi:10.1039/C5NR06873H

[25] Banszerus L, Schmitz M, Engels S, Dauber J, Oellers M, Haupt F, Watanabe K, Taniguchi T, Beschoten B, Stampfer C. Ultrahigh-mobility graphene devices from chemical vapor deposition on reusable copper. Science Advance. 2015;**1**:1–6. doi:10.1126/sciadv.1500222

[26] Xu SC, Man BY, Jiang SZ, Chen CS, Yang C, Liu M, Gao XG, Sun ZC, Zhang C. Direct synthesis of graphene on SiO2 substrates by chemical vapor deposition. CrystEngComm. 2013;**15**:1840. doi:10.1039/c3ce27029g

[27] Kim H, Song I, Park C, Son M, Hong M, Kim Y, Kim JS, Shin HJ, Baik J, Choi HC. Copper-vapor-assisted chemical vapor deposition for high-quality and metal-free single-layer graphene on amorphous SiO2 substrate. ACS Nano. 2013;**7**:6575–6582. doi:10.1021/nn402847w

[28] McNerny DQ, Viswanath B, Copic D, Laye FR, Prohoda C, Brieland-Shoultz AC, Polsen ES, Dee NT, Veerasamy VS, Hart AJ. Direct fabrication of graphene on SiO2 enabled by thin film stress engineering. Scientific Reports. 2014;**4**:5049. doi:10.1038/srep05049

[29] Shautsova V, Gilbertson AM, Black NCG, Maier SA, Cohen LF. Hexagonal boron nitride assisted transfer and encapsulation of large area CVD graphene. Scientific Reports. 2016;**6**:30210. doi:10.1038/srep30210

[30] Li Q, Liu M, Zhang Y, Liu Z. Hexagonal boron nitride-graphene heterostructures: Synthesis and interfacial properties. Small. 2016;**12**:32–50. doi:10.1002/smll.201501766

[31] Sun J, Lindvall N, Cole MT, Teo KBK, Yurgens A. Large-area uniform graphene-like thin films grown by chemical vapor deposition directly on silicon nitride. Applied Physics Letters. 2011;**98**. doi:10.1063/1.3602921

[32] Schmid S, Bagci T, Zeuthen E, Taylor JM, Herring PK, Cassidy MC, Marcus CM, Guillermo Villanueva L, Amato B, Boisen A, Shin YC, Kong J, Sørensen AS, Usami K, Polzik ES. Single-layer graphene on silicon nitride micromembrane resonators. Journal of Applied Physics. 2014;**115**. doi:10.1063/1.4862296

[33] Iyer GRS, Wang J, Wells G, Bradley MP, Borondics F. Nanoscale imaging of freestanding nitrogen doped single layer graphene. Nanoscale. 2015;**7**:2289–2294. doi:10.1039/c4nr05385k

[34] Zhang M, Hou C, Halder A, Ulstrup J, Chi Q. Interlocked graphene Prussian blue hybrid composites enable multifunctional electrochemical applications. Biosensors and Bioelectronics. 2017;**89**:570–577. doi:10.1016/j.bios.2016.02.044

[35] Shao Y, El-Kady MF, Lin CW, Zhu G, Marsh KL, Hwang JY, Zhang Q, Li Y, Wang H, Kaner RB. 3D Freeze-casting of cellular graphene films for ultrahigh-power-density supercapacitors. Advanced Materials. 2016:6719–6726. doi:10.1002/adma.201506157

[36] Zhu N, Han S, Gan S, Ulstrup J, Chi Q. Graphene paper doped with chemically compatible Prussian Blue nanoparticles as nanohybrid electrocatalyst. Advanced Functional Material. 2013;**23**:5297–5306. doi:10.1002/adfm.201300605

[37] Zhang M, Halder A, Hou C, Ulstrup J, Chi Q. Free-standing and flexible graphene papers as disposable non-enzymatic electrochemical sensors. Bioelectrochemistry. 2016;**109**:87–94. doi:10.1016/j.bioelechem.2016.02.002

[38] Choi W, Choi J, Bang J, Lee JH. Layer-by-layer assembly of graphene oxide nanosheets on polyamide membranes for durable reverse-osmosis applications. ACS Applied Material Interfaces. 2013;**5**:12510–12519. doi:10.1021/am403790s

[39] Lin X, Jia J, Yousefi N, Shen X, Kim J-K. Excellent optoelectrical properties of graphene oxide thin films deposited on a flexible substrate by Langmuir-Blodgett assembly. Journal of Material Chemistry C. 2013;**1**:6869–6877. doi:10.1039/c3tc31497a

[40] Zheng Q, Ip WH, Lin X, Yousefi N, Yeung KK, Li Z, Kim JK. Transparent conductive films consisting of ultralarge graphene sheets produced by Langmuir-Blodgett assembly. ACS Nano. 2011;**5**:6039–6051. doi:10.1021/nn2018683

[41] Zou J, Kim F. Diffusion driven layer-by-layer assembly of graphene oxide nanosheets into porous three-dimensional macrostructures. Nature Communication. 2014;**5**:5254. doi:10.1038/ncomms6254

[42] Dai M-K, Lian J-T, Lin T-Y, Chen Y-F. High-performance transparent and flexible inorganic thin film transistors: a facile integration of graphene nanosheets and amorphous InGaZnO. Journal of Material Chemstry C. 2013;**1**:5064–5071. doi:10.1039/c3tc30890a

[43] Bae SH, Kahya O, Sharma BK, Kwon J, Cho HJ, Özyilmaz B, Ahn JH. Graphene-P(VDF-TrFE) multilayer film for flexible applications. ACS Nano. 2013;**7**:3130–3138. doi:10.1021/nn400848j

[44] Borini S, White R, Wei D, Astley M, Haque S, Spigone E, Harris N, Kivioja J, Ryhänen T. Ultrafast graphene oxide humidity sensors ACS Nano. 2013;**7**:11166–11173. doi:10.1021/nn404889b

[45] Bin Yao H, Ge J, Wang CF, Wang X, Hu W, Zheng ZJ, Ni Y, Yu SH. A flexible and highly pressure-sensitive graphene-polyurethane sponge based on fractured microstructure design. Advanced Material. 2013;**25**:6692–6698. doi:10.1002/adma.201303041

[46] Seo S, Min M, Lee J, Lee T, Choi SY, Lee H. Solution-processed reduced graphene oxide films as electronic contacts for molecular monolayer junctions. Angewendte Chemie International Edition. 2012;**51**:108–112. doi:10.1002/anie.201105895

[47] Li X, Zhang G, Bai X, Sun X, Wang X, Wang E, Dai H. Highly conducting graphene sheets and Langmuir-Blodgett films. Nature Nanotechnology. 2008;**3**:538–42. doi:10.1038/nnano.2008.210

[48] Li T, Hauptmann JR, Wei Z, Petersen S, Bovet N, Vosch T, Nygãrd J, Hu W, Liu Y, Bjørnholm T, Nørgaard K, Laursen BW. Solution-processed ultrathin chemically derived graphene films as soft top contacts for solid-state molecular electronic junctions. Advanced Material. 2012;**24**:1333–1339. doi:10.1002/adma.201104550

[49] Wee BH, Hong JD. A method for fabricating an ultrathin multilayer film composed of poly(p-phenylenevinylene) and reduced graphene oxide on a plastic substrate for flexible optoelectronic applications. Advanced Functional Material. 2013;**23**:4657–4666. doi:10.1002/adfm.201300224

[50] Sha MSP, Boccaccini AR. Applications of graphene electrophoretic deposition. A Review. Physical Chemistry B. 2012;**117**:1502–1515. doi:org/10.1021/jp3064917

[51] Zhang M, Hou C, Halder A, Wang H, Chi Q. Graphene papers: smart architecture and specific functionalization for biomimetics, electrocatalytic sensing and energy storage. Materials Chemistry Frontiers. 2017;**1**:37–60. doi:10.1039/C6QM00145A

[52] Anis A, Mohammad AA, Hussain A, Ahmed SE. Reduced graphene oxide thin film on conductive substrates by bipolar electrochemistry. Scientific Reports. 2016;**6**:21282. doi:10.1038/srep21282

[53] Cao X, Qi D, Yin S, Bu J, Li F, Goh CF, Zhang S, Chen X. Ambient fabrication of large-area graphene films via a synchronous reduction and assembly strategy. Advanced Materials. 2013;**25**:2957–2962. doi:10.1002/adma.201300586

[54] Yin S, Goldovsky Y, Herzberg M, Liu L, Sun H, Zhang Y, Meng F, Cao X, Sun DD, Chen H, Kushmaro A, Chen X. Functional free-standing graphene honeycomb films. Advanced Functional Materials. 2013;**23**:2972–2978. doi:10.1002/adfm.201203491

[55] Yin S, Zhang Y, Kong J, Zou C, Li CM, Lu X, Ma J, Boey FYC, Chen X. Assembly of graphene sheets into hierarchical structures for high-performance energy storage. ACS Nano. 2011;**5**:3831–3838. doi:10.1021/nn2001728

[56] Chen C-M, Zhang Q, Huang C-H, Zhao X-C, Zhang B-S, Kong Q-Q, Wang M-Z, Yang Y-G, Cai R, Sheng Su D. Macroporous "bubble" graphene film via template-directed ordered-assembly for high rate supercapacitors. Chemical Communication. 2012;**48**:7149. doi:10.1039/c2cc32189k

[57] Ping H, Zhang M, Li H, Li S, Chen Q, Sun C, Zhang T. Visual detection of melamine in raw milk by label-free silver nanoparticles. Food Control. 2012;**23**:191–197. doi:10.1016/j.foodcont.2011.07.009

[58] Zhang M, Cao X, Li H, Guan F, Guo J, Shen F, Luo Y, Sun C, Zhang L. Sensitive fluorescent detection of melamine in raw milk based on the inner filter effect of Au nanoparticles on the fluorescence of CdTe quantum dots. Food Chemistry. 2012;**135**:1894–1900. doi:10.1016/j.foodchem.2012.06.070

[59] Yun M, Choe JE, You JM, Ahmed MS, Lee K, Üstündağ Z, Jeon S. IGH catalytic activity of electrochemically reduced graphene composite toward electrochemical sensing of Orange II. Food Chemistry. 2015;**169**:114–119. doi:10.1016/j.foodchem.2014.07.143

[60] Xiao F, Song J, Gao H, Zan X, Xu R, Duan H. Coating graphene paper with 2D-assembly of electrocatalytic nanoparticles: A modular approach toward high-performance flexible electrodes. ACS Nano. 2012;**6**:100–110. doi:10.1021/nn202930m

[61] Xiao F, Li Y, Gao H, Ge S, Duan H. Growth of coral-like PtAu-MnO2 binary nanocomposites on free-standing graphene paper for flexible nonenzymatic glucose sensors. Biosensors and Bioelectronics. 2013;**41**:417–423. doi:10.1016/j.bios.2012.08.062

[62] Xi Q, Chen X, Evans DG, Yang W. Gold nanoparticle-embedded porous graphene thin films fabricated via layer-by-layer self-assembly and subsequent thermal annealing for electrochemical sensing. Langmuir. 2012;**28**:9885–9892. doi:10.1021/la301440k

[63] Zhong L, Gan S, Fu X, Li F, Han D, Guo L, Niu L. Electrochemically controlled growth of silver nanocrystals on graphene thin film and applications for efficient nonenzymatic H_2O_2 biosensor. Electrochimica Acta. 2013;**89**:222–228. doi:10.1016/j.electacta.2012.10.161

[64] Kholmanov IN, Magnuson CW, Aliev AE, Li H, Zhang B, Suk JW, Zhang LL, Peng E, Mousavi SH, Khanikaev AB, Piner R, Shvets G, Ruoff RS. Improved electrical conductivity of graphene films integrated with metal nanowires. Nano Letters. 2012;**12**:5679–5683. doi:10.1021/nl302870x

[65] Du Y, Zhao Y, Qu Y, Chen C-H, Chen C-M, Chuang C-H, Zhu Y. Enhanced light-matter interaction of graphene-gold nanoparticle hybrid films for high-performance SERS detection. Journal of Material Chemistry C. 2014;**2**:4683–4691. doi:10.1039/c4tc00353e

[66] Liu Y, Chang Q, Huang L. Transparent, flexible conducting graphene hybrid films with a subpercolating network of silver nanowires. Journal of Material Chemistry C. 2013;**1**:2970. doi:10.1039/c3tc30178h

[67] Mulpur P, Podila R, Lingam K, Vemula SK, Ramamurthy SS, Kamisetti V, Rao AM. Amplification of surface plasmon coupled emission from graphene-ag hybrid films. Journal of Physics and Chemistry C. 2013;**117**:17205–17210. doi:10.1021/jp406122s

[68] Dao TD, Hong JE, Ryu KS, Jeong HM. Super-tough functionalized graphene paper as a high-capacity anode for lithium ion batteries. Chemical Engineering Journal. 2014;**250**: 257–266. doi:10.1016/j.cej.2014.04.051

[69] Zhou Y, Yang J, Cheng X, Zhao N, Sun H, Li D. Transparent and conductive reduced graphene oxide/silver nanoparticles multilayer film obtained by electrical self-assembly process with graphene oxide sheets and silver colloid. RSC Advances. 2013;**3**:3391. doi:10.1039/c2ra22256f

[70] Liu F, Piao Y, Choi KS, Seo TS. Fabrication of free-standing graphene composite films as electrochemical biosensors. Carbon. 2012;**50**:123–133. doi:10.1016/j.carbon.2011.07.061

[71] Zhang P, Zhang X, Zhang S, Lu X, Li Q, Su Z, Wei G. One-pot green synthesis, characterizations, and biosensor application of self-assembled reduced graphene oxide–gold nanoparticle hybrid membranes. Journal of Materials Chemistry B. 2013;**1**:6525. doi:10.1039/c3tb21270j

[72] Zhang H, Zhang X, Zhang D, Sun X, Lin H, Wang C, Ma Y. One-step electrophoretic deposition of reduced graphene oxide and $Ni(OH)_2$ composite films for controlled syntheses supercapacitor electrodes. Journal of Physics and Chemistry B.2013;**117**:1616–1627. doi:10.1021/jp305198j

[73] Chen S, Qiao SZ. Hierarchically porous nitrogen-doped graphene-$NiCo_2O_4$ hybrid paper as an advanced electrocatalytic water-splitting material. ACS Nano. 2013;**7**:10190–10196. doi:10.1021/nn404444r

[74] Wang X, Cao X, Bourgeois L, Guan H, Chen S, Zhong Y, Tang DM, Li H, Zhai T, Li L, Bando Y, Golberg D. N-doped graphene-SnO_2 sandwich paper for high-performance lithium-ion batteries. Advanced Functional Materials. 2012;**22**:2682–2690. doi:10.1002/adfm.201103110

[75] Chen D, Tang L, Li J. Graphene-based materials in electrochemistry. Chemical Society Reviews. 2010;**39**:3157. doi:10.1039/b923596e

[76] Zhng M, Hou C, Halder A, Chi Q. Ultralight, flexible and semi-transparent metal oxide papers for photoelectrochemical water splitting. ACS Applied Materials Interfaces. 2017;**9**:3922–3930. doi:10.1021/acsami.6b14036

[77] Zhu X, Zhu Y, Murali S, Stoller MD, Ruoff RS. Nanostructured reduced graphene oxide/Fe_2O_3 composite as a high-performance anode material for lithium ion batteries. ACS Nano. 2011;**5**:3333–3338. doi:10.1021/nn200493r

[78] Zhu H, Lee KT, Hitz GT, Han X, Li Y, Wan J, Lacey S, Cresce AVW, Xu K, Wachsman E, Hu L. Free-standing $Na_{2/3}Fe_{1/2}Mn_{1/2}O_2$@Graphene film for a sodium-ion battery cathode. ACS Applied Materials Interfaces. 2014;**6**:4242–4247. doi:10.1021/am405970s

[79] Xiao F, Li Y, Zan X, Liao K, Xu R, Duan H. Growth of metal-metal oxide nanostructures on freestanding graphene paper for flexible biosensors. Advanced Functional Materials. 2012;**22**:2487–2494. doi:10.1002/adfm.201200191

[80] Sumboja A, Foo CY, Wang X, Lee PS. Large areal mass, flexible and free-standing reduced graphene oxide/manganese dioxide paper for asymmetric supercapacitor device. Advanced Materials. 2013;**25**:2809–2815. doi:10.1002/adma.201205064

[81] Cheng J, Wang B, Xin HL, Yang G, Cai H, Nie F, Huang H. Self-assembled V_2O_5 nanosheets/reduced graphene oxide hierarchical nanocomposite as a high-performance cathode material for lithium ion batteries. Journal of Materials Chemistry A. 2013;**1**:10814. doi:10.1039/c3ta12066j

[82] Zou Y, Wang Y. NiO nanosheets grown on graphene nanosheets as superior anode materials for Li-ion batteries. Nanoscale. 2011;**3**:2615–2620. doi:10.1039/c1nr10070j

[83] Gan T, Sun J, Meng W, Song L, Zhang Y. Electrochemical sensor based on graphene and mesoporous TiO_2 for the simultaneous determination of trace colourants in food. Food Chemistry. 2013;**141**:3731–3737. doi:10.1016/j.foodchem.2013.06.084

[84] Wang D, Choi D, Li J, Yang Z, Nie Z, Kou R, Hu D, Wang C, Saraf LV, Zhang J, Aksay IA, Liu J. Self-assembled TiO_2 – graphene hybrid nanostructures for enhanced Li-ion insertion. ACS Nano. 2009;**3**:907–914. doi:10.1021/nn900150y

[85] Fan Y, Huang K-J, Niu D-J, Yang C-P, Jing Q-S. TiO_2-graphene nanocomposite for electrochemical sensing of adenine and guanine. Electrochimica Acta. 2011;**56**:4685–4690. doi:10.1016/j.electacta.2011.02.114

[86] Lee CH, Kim YJ, Hong YJ, Jeon SR, Bae S, Hong BH, Yi GC. Flexible inorganic nanostructure light-emitting diodes fabricated on graphene films. Advanced Materials. 2011;**23**:4614–4619. doi:10.1002/adma.201102407

[87] Feng XM, Li RM, Ma YW, Chen RF, Shi NE, Fan QL, Huang W. One-step electrochemical synthesis of graphene/polyaniline composite film and its applications. Advanced Functional Materials. 2011;**21**:2989–2996. doi:10.1002/adfm.201100038

[88] Kong FY, Gu SX, Li WW, Chen TT, Xu Q, Wang W. A paper disk equipped with graphene/polyaniline/Au nanoparticles/glucose oxidase biocomposite modified screen-printed electrode: Toward whole blood glucose determination. Biosensors and Bioelectronics. 2014;**56**:77–82. doi:10.1016/j.bios.2013.12.067

[89] Chi K, Zhang Z, Xi J, Huang Y, Xiao F, Wang S, Liu Y. Freestanding graphene paper supported three-dimensional porous graphene-polyaniline nanocomposite synthesized by inkjet printing and in flexible all-solid-state supercapacitor. ACS Applied Materials Interfaces. 2014;**6**:16312–16319. doi:10.1021/am504539k

[90] Cong HP, Ren XC, Wang P, Yu SH. Flexible graphene-polyaniline composite paper for high-performance supercapacitor. Energy Environmental Science. 2013;**6**:1185–1191. doi:10.1039/c2ee24203f

[91] Xiang J, Drzal LT. Improving thermoelectric properties of graphene/polyaniline paper by folding. Chemical Physics Letters. 2014;**593**:109–114. doi:10.1016/j.cplett.2013.12.079

[92] Xu Y, Lou B, Lv Z, Zhou Z, Zhang L, Wang E. Paper-based solid-state electrochemiluminescence sensor using poly(sodium 4-styrenesulfonate) functionalized graphene/nafion composite film. Analytica Chimica Acta. 2013;**763**:20–27. doi:10.1016/j.aca.2012.12.009

[93] Mu J, Hou C, Zhu B, Wang H, Li Y, Zhang Q. A multi-responsive water-driven actuator with instant and powerful performance for versatile applications. Scientific Reports. 2015;**5**:9503. doi:10.1038/srep09503

[94] Mu J, Hou C, Wang H, Li Y, Zhang Q, Zhu M. Origami-inspired active graphene-based paper for programmable instant self-folding walking devices. Scientific Advances. 2015;**1**

[95] Halder A, Zhang M, Chi Q. Electroactive and biocompatible functionalization of graphene for the development of biosensing platforms. Biosensors and Bioelectronics. 2017;**87**:764–771. doi:10.1016/j.bios.2016.09.030

[96] Zhao X, Zhang P, Chen Y, Su Z, Wei G. Recent advances in the fabrication and structure-specific applications of graphene-based inorganic hybrid membranes. Nanoscale. 2015;**7**:5080–5093. doi:10.1039/C5NR00084J

[97] Dinesh B, Mani V, Saraswathi R, Chen S-M. Direct electrochemistry of cytochrome c immobilized on a graphene oxide–carbon nanotube composite for picomolar detection of hydrogen peroxide. RSC Advances. 2014;**4**:28229. doi:10.1039/c4ra02789b

[98] Ensafi AA, Jafari-Asl M, Dorostkar N, Ghiaci M, Martinez-Huerta MV, Fierro JLG. The fabrication and characterization of Cu-nanoparticle immobilization on a hybrid chitosan derivative-carbon support as a novel electrochemical sensor: Application for the sensitive enzymeless oxidation of glucose and reduction of hydrogen peroxide. Journal of Materials Chemistry B. 2014;**2**:706–717. doi:10.1039/C3TB21434F

[99] Sun Z, Fu H, Deng L, Wang J. Redox-active thionine-graphene oxide hybrid nanosheet: One-pot, rapid synthesis, and application as a sensing platform for uric acid. Analytica Chimica Acta. 2013;**761**:84–91. doi:10.1016/j.aca.2012.11.057

[100] Luo L, Zhu L, Wang Z. Nonenzymatic amperometric determination of glucose by CuO nanocubes-graphene nanocomposite modified electrode. Bioelectrochemistry. 2012;**88**:156–163. doi:10.1016/j.bioelechem.2012.03.006

[101] Tian Y, Liu Y, Wang WP, Zhang X, Peng W. CuO nanoparticles on sulfur-doped graphene for nonenzymatic glucose sensing. Electrochimica Acta. 2015;**156**:244–251. doi:10.1016/j.electacta.2015.01.016

[102] Zhu X, Jiao Q, Zhang C, Zuo X, Xiao X, Liang Y, Nan J. Amperometric nonenzymatic determination of glucose based on a glassy carbon electrode modified with nickel(II) oxides and graphene. Microchimica Acta. 2013;**180**:477–483. doi:10.1007/s00604-013-0955-1

[103] Ci S, Mao S, Huang T, Wen Z, Steeber DA, Chen J. Enzymeless glucose detection based on CoO/graphene microsphere hybrids. Electroanalysis. 2014;**26**:1326–1334. doi:10.1002/elan.201300645

[104] Pandikumar A, Soon How GT, See TP, Omar FS, Jayabal S, Kamali KZ, Yusoff N, Jamil A, Ramaraj R, John SA, Lim HN, Huang NM. Graphene and its nanocomposite material based electrochemical sensor platform for dopamine. RSC Advance. 2014;**4**: 63296–63323. doi:10.1039/C4RA13777A

[105] Sheng Z-H, Zheng X-Q, Xu J-Y, Bao W-J, Wang F-B, Xia X-H. Electrochemical sensor based on nitrogen doped graphene: Simultaneous determination of ascorbic acid, dopamine and uric acid. Biosensors and Bioelectronics. 2012;**34**:125–131. doi:10.1016/j.bios.2012.01.030

[106] Dong X, Wang X, Wang L, Song H, Zhang H, Huang W, Chen P. 3D graphene foam as a monolithic and macroporous carbon electrode for electrochemical sensing. ACS Applied Materials Interfaces. 2012;**4**:3129–3133. doi:10.1021/am300459m

[107] Li M, Liu C, Zhao H, An H, Cao H, Zhang Y, Fan Z. Tuning sulfur doping in graphene for highly sensitive dopamine biosensors. Carbon New York. 2015;**86**:197–206. doi:10.1016/j.carbon.2015.01.029

[108] Yu B, Kuang D, Liu S, Liu C, Zhang T. Template-assisted self-assembly method to prepare three-dimensional reduced graphene oxide for dopamine sensing. Sensors Actuators, B Chemistry. 2014;**205**:120–126. doi:10.1016/j.snb.2014.08.038

[109] Yu X, Sheng K, Shi G. A three-dimensional interpenetrating electrode of reduced graphene oxide for selective detection of dopamine. Analyst. 2014;**139**:4525–4531. doi:10.1039/c4an00604f

[110] Bagherzadeh M, Heydari M. Electrochemical detection of dopamine based on preconcentration by graphene nanosheets. Analyst. 2013;**138**:6044–51. doi:10.1039/c3an01318a

[111] Barsan MM, David M, Florescu M, Ţugulea L, Brett CMA. A new self-assembled layer-by-layer glucose biosensor based on chitosan biopolymer entrapped enzyme with nitrogen doped graphene. Bioelectrochemistry. 2014;**99**:46–52. doi:10.1016/j.bioelechem.2014.06.004

[112] Cao X, Ye Y, Li Y, Xu X, Yu J, Liu S. Self-assembled glucose oxidase/graphene/gold ternary nanocomposites for direct electrochemistry and electrocatalysis. Journal of Electroanalytical Chemistry. 2013;**697**:10–14. doi:10.1016/j.jelechem.2013.03.001

[113] Cai C-J, Xu M-W, Bao S-J, Lei C, Jia D-Z. A facile route for constructing a graphene-chitosan-ZrO2 composite for direct electron transfer and glucose sensing. RSC Advance. 2012;**2**:8172–8178. doi:10.1039/c2ra20926h

[114] Zhong X, Yuan R, Chai YQ. Synthesis of chitosan-Prussian blue-graphene composite nanosheets for electrochemical detection of glucose based on pseudobienzyme channeling. Sensors Actuators, B Chemistry. 2012;**162**:334–340. doi:10.1016/j.snb.2011.12.091

[115] Qiu J-D, Huang J, Liang R-P. Nanocomposite film based on graphene oxide for high performance flexible glucose biosensor. Sensors Actuators B Chemistry. 2011;**160**:287–294. doi:10.1016/j.snb.2011.07.049

[116] Peng H, Huang Z, Zheng Y, Chen W, Liu A, Lin X. A novel nanocomposite matrix based on graphene oxide and ferrocene-branched organically modified sol-gel/chitosan for biosensor application. Journal of Solid State Electrochemistry. 2014;**18**:1941–1949. doi:10.1007/s10008-014-2415-1

[117] Wang Z, Hu Y, Yang W, Zhou M, Hu X. Facile one-step microwave-assisted route towards Ni nanospheres/reduced graphene oxide hybrids for non-enzymatic glucose sensing. Sensors. 2012;**12**:4860–4869. doi:10.3390/s120404860

[118] Zhang Y, Xiao X, Sun Y, Shi Y, Dai H, Ni P, Hu J, Li Z, Song Y, Wang L. Electrochemical deposition of nickel nanoparticles on reduced graphene oxide film for nonenzymatic glucose sensing. Electroanalysis. 2013;**25**:959–966. doi:10.1002/elan.201200479

[119] Liang B, Fang L, Hu Y, Yang G, Zhu Q, Ye X. Fabrication and application of flexible graphene silk composite film electrodes decorated with spiky Pt nanospheres. Nanoscale. 2014;**6**:4264–4274. doi:10.1039/c3nr06057h

[120] Hsu YW, Hsu TK, Sun CL, Nien YT, Pu NW, Der Ger M. Synthesis of CuO/graphene nanocomposites for nonenzymatic electrochemical glucose biosensor applications. Electrochimica Acta. 2012;**82**:152–157. doi:10.1016/j.electacta.2012.03.094

[121] Dong XC, Xu H, Wang XW, Huang YX, Chan-Park MB, Zhang H, Wang LH, Huang W, Chen P. 3D graphene-cobalt oxide electrode for high-performance supercapacitor and enzymeless glucose detection. ACS Nano. 2012;**6**:3206–3213. doi:10.1021/nn300097q

[122] Gao W, Chen Y, Xi J, Lin S, Chen Y, Lin Y, Chen Z. A novel electrochemiluminescence ethanol biosensor based on tris(2,2′-bipyridine) ruthenium (II) and alcohol dehydrogenase immobilized in graphene/bovine serum albumin composite film. Biosensors and Bioelectronics. 2013;**41**:776–782. doi:10.1016/j.bios.2012.10.005

[123] Li J, Li X, Zhang Y, Li R, Wu D, Du B, Zhang Y, Ma H, Wei Q. Electrochemiluminescence sensor based on cationic polythiophene derivative and NH 2 –graphene for dopamine detection. RSC Advance. 2015;**5**:5432–5437. doi:10.1039/C4RA14595J

[124] Xu Y, Lv Z, Xia Y, Han Y, Lou B, Wang E. Highly porous magnetite/graphene nanocomposites for a solid-state electrochemiluminescence sensor on paper-based chips. Analytical and Bioanalytical Chemistry. 2013;**405**:3549–3558. doi:10.1007/s00216-012-6510-9

[125] Delle LE, Huck C, Bäcker M, Müller F, Grandthyll S, Jacobs K, Lilischkis R, Vu XT, Schöning MJ, Wagner P, Thoelen R, Weil M, Ingebrandt S. Impedimetric immunosensor for the detection of histamine based on reduced graphene oxide. Physica Status Solidi Applied Materials Science. 2015;**212**:1327–1334. doi:10.1002/pssa.201431863

[126] Loo AH, Bonanni A, Ambrosi A, Poh HL, Pumera M. Impedimetric immunoglobulin G immunosensor based on chemically modified graphenes. Nanoscale. 2012;**4**:921–925. doi:10.1039/c2nr11492e

[127] Wang Y, Ping J, Ye Z, Wu J, Ying Y. Impedimetric immunosensor based on gold nanoparticles modified graphene paper for label-free detection of Escherichia coli O157:H7. Biosensors and Bioelectronics. 2013;**49**:492–8. doi:10.1016/j.bios.2013.05.061

Section 3

State-of-the-Art Gas Sensors

Chapter 4

Electrochemical Sensors for Monitoring of Indoor and Outdoor Air Pollution

Igor Cretescu, Doina Lutic and
Liliana Rosemarie Manea

Additional information is available at the end of the chapter

http://dx.doi.org/10.5772/intechopen.68512

Abstract

This chapter aims a comprehensive presentation of the most common electrochemical sensors used in the real monitoring applications of air purity testing. Oxygen, hydrogen, hydrogen sulfide, nitrogen oxides, carbon monoxide and carbon dioxide are gases, which can be accurately detected and measured. Too high or too low oxygen concentration levels make the air improper for breathing. Hydrogen sulfide and carbon monoxide are dangerous species; any leakage needs to be pinpointed. A calibrated network of sensors for monitoring gas detection makes it possible to easily locate the source of gas escape during indoor air monitoring. The air quality monitoring stations based on electrochemical sensors are nowadays used to determine the global pollution index of the atmospheric air, in order to prevent the risks toward the human health and damage of environment, especially in the highly populated and industrialized urban areas.

Keywords: electrochemical sensors, sensing mechanism, semiconductive oxides, sensing surface, nanoparticles

1. Introduction

Some dramatic changes in the atmospheric chemistry occurred in the last few decades in tight connection with the intense economic overall development of the human society. The highest advances are noticed in the field of energy, transportation, communications and production of numerous daily commodities as cosmetics, hygiene and care products, and pharmaceuticals. This progress had to pay a high price on the quality of the environment in terms of atmosphere, soil and water pollution, climate changes induced by human activities, deforestation and contribution to global warming.

The detection of various gas species with polluting effects as carbon monoxide, hydrogen sulfide, nitrogen and sulfur oxide, ozone, in both indoor and outdoor air, is a challenging priority worldwide, due to their complex and destructive effects on the environment. Even low concentration values of these species are associated with risks toward human health and generalized damage to ecosystems. Therefore, the detection of gases using electrochemical sensors earns continuous popularity. "Four S" requirements can be formulated and related to chemical sensing: sensitivity, selectivity, stability and speed of response. Nowadays, sensors also bring high reproducibility, robustness, relatively simple setup and involve low electricity consummation for the detection of various gaseous species and concentrations over a wide range of operation conditions, for indoor and outdoor air [1].

On an electrochemical gas sensor surface, the detected molecules undergo a redox reaction on a suitable electrode, generating an electrical current depending on the gas concentration. The sensitive layer from the sensing surface plays the role of a *receptor*. Specific functional surface groups allow the adsorption or chemosorption of the detected species, sometimes followed by combustion or other chemical reactions involving electron transfer. The receptor is connected to a *transducer*, a device which transforms the atomic scale interaction to an electrical measurable response. The changes of the physical properties of the receptor surface due to the interaction with the sensed molecules refer to electron density, optical properties and mass or temperature. The receptor of most electrochemical sensor systems is based on semiconductive metallic oxides (SMO) and the changes during the interaction of a gas involve changes of the electronic charge density and charge carrier mobility [2].

The complex requirements connected to the need of fast and reliable measurement of various gas species, using portable devices if possible, switched the designers and manufacturers interest toward miniaturized setups, usually embedded in so-called sensor arrays. An important advantage of these devices is their extremely high selectivity, possible to reach by performing the signal processing with appropriate software. These advances became possible due to the application of advanced microelectronic and microelectromechanical systems (MEMS) processing.

Different nanomaterials (mainly nanoparticles and nanowires) have also been used in the last few decades in the formulation of sensing layers for the electrochemical sensors. These structures have, on one part, high surface areas and, on another part, the nanosize of the particles or wires from the sensing structure, which makes it possible to be a special matching between the surface and the adsorbed chemical analytes. This fact makes it possible to be used for the manufacturing of sensors with high improvement of the sensitivity level.

A sensor's response to a chemical compound is usually expressed by comparing the signal in the presence of an analyte with the signal in pure air or in an inert gas. The response can thus be the difference or the ratio between the signal values in the absence and presence of the sensed species, as well as the derivative or integrative change with time of the initial signal associated with the presence of a chemical species [2, 3].

In this chapter, a brief description will be given about the sensing mechanisms; then the most popular materials used in the chemical sensing, outlining the role of the nanoporous structure of the sensing layer will be reviewed, and finally, some examples of practical applications of chemical sensors for dangerous gases will be displayed.

2. The principles of gas sensing

During the detection process, some physical or chemical interactions occur between the measured gas and certain specific sites from the surface of the sensor. Sometimes, the changes are even deeper, so the electronic structure of the bulk phase of the sensing material is also affected. In most cases, while the gas interacts with specific sites from the sensing surface, the electronic structure of the solid is highly perturbed and can be converted to measurable electrical properties. It is important to mention, especially for the sensors based on sensing nanostructures (nanoparticles, nanowires, quantum dots), that the interactions take place at atomic level. The sensing layer, which is in direct contact with the gas, is called *receptor*. The interaction between the receptor and the gas induces several changes in measurable parameters of the layer as electrical charge, polarizability, electrical potential or initiating heat generation, changes in mass (detected by measuring the resonating frequency) or optical properties. The signal from the receptor is transformed in a measurable output by a *transducer*. The transducer's role highlights the importance of the receptor microstructure, since the packing extent of the particles has an overwhelming influence on the quality of the signal.

The physical parameters used as output signal are very diverse and depend on the sensed species and receptor composition, transducer's detailed structure, detected species concentration range, required sensor accuracy in detecting and/or measuring, detailed design of the sensing device, working conditions, possible interfering between the target species and other gases from the environment, allowed energy consummation, etc.

A classification of the types of sensor systems, based on the sensing mechanisms, indicates the following categories [2]:

1. Measuring the electrical resistance or impedance changes occurring on thick film metals or metal oxides or conducting polymers, printed on interdigitated electrodes;
2. Amperometric and potentiometric sensors, based on electrical current changes;
3. Capacitive sensors, measuring the electrical capacitance of the layer of metal oxide semiconductor;
4. Field effect transistors (FETs), based on the changes of the work function or polarization of the layer;
5. Acoustic wave resonators, by measuring the mass changes of the bulk or of the surface;
6. Pellistors, thermistors or thermopiles, based on temperature changes;
7. Optical sensors, following the changes of the optical adsorption or reflection due to changes in refractive index and optical layer thickness;

8. Surface plasmon resonance (SPR) sensors, measuring changes in the incidence angle for the start of a surface plasmon, due to changes in refractive index.

The most common parameter modified during the gas exposure of the sensing layer is the value of electrical resistance of the layer. The reason for this behavior is the chemical reaction between a reductive gas species, chemisorbed on the sensor surface, and oxygen from its network. Typically, tin or titanium oxides are appropriate providers of oxygen. The sensors are effective at relatively high temperatures (150–500°C), where the semiconductive oxides become quite good conductors. The reductive gas could be hydrogen or carbon monoxide.

$$CO + O^{-}{}_{ad}(SnO_{2-X}) \rightarrow CO_2 + (SnO_{2-X})^{+} \qquad (r-1)$$

$$1/2O_2 + (SnO_{2-X}) \rightarrow O^{-}{}_{ad}(SnO_{2-X}) \qquad (r-2)$$

When the combustible (reducing) gas consumes the oxygen from the semiconductive material, the amount of oxygen ions from the oxide network decreases and so does the value of the electrical resistance. As soon as the environment of the sensor gets enriched in oxygen, it oxidizes the SnO to SnO_2 again (**Figure 1**).

The electrochemical cells, operating in the potentiometric or amperometric mode, are nowadays the most common gas sensors.

In both cases, the gas diffuses to the working electrode to the sensing area of the sensor through a porous membrane. On the electrode, the gas reacts by oxidation or reduction reactions, giving rise to an electric current, which is taken as a signal by the external circuit. In case of amperometric sensors (**Figure 2**) (two electrode sensor), the measured signal is amplified and further processed, while maintaining a voltage between the working and counter electrodes. In case of potentiometric sensors, the work function requires a three-electrode sensor and the voltage is kept between the working and reference electrodes for a three electrode cell (**Figure 2**). The oxidation and reduction reactions occur in opposite pairs, so if at the working electrode takes place an oxidation; at the counter electrode will take place a reduction.

The amperometric sensors are the most used electrochemical gas sensors, due to linear dependence of the generated current with the gas concentration. In the electrochemical cell, the equilibrium is not established, the current being related to the rate of the electrolytic process occurring at the sensing electrode (named working electrode), whose potential is kept constant using another electrode (named reference electrode). The third electrode is also called counter electrode and is included in an electrochemical sensor (the cell is also called three-electrode sensor). Therefore, the operation of an electrochemical sensor requires a potentiostatic circuit.

The value of the developed current is proportional to the amount of the target gas oxidized on the working electrode. The design of sensors is usually manufactured so that the gas supply to the sensing area is limited by diffusion, in order to obtain a sensor signal linearly proportional to the gas concentration. The signal linearity is an important advantage of the electrochemical sensors compared to other sensor types, in which case, the signal must be linearized. Together with the simplicity of signal interpretation, the linear output allows making more

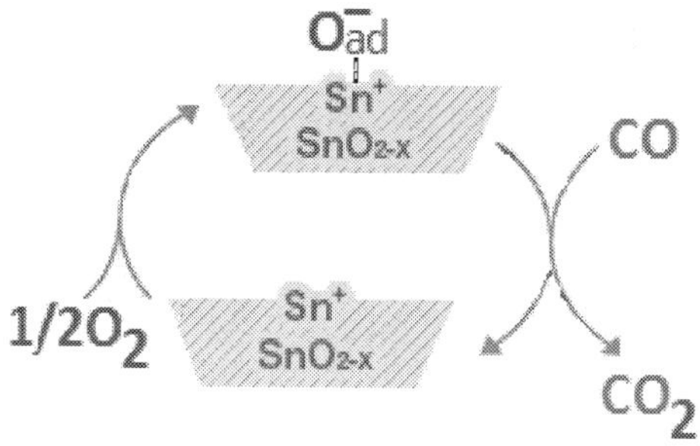

Figure 1. Scheme of the reaction between CO and adsorbed oxygen O_{ad} on the SnO_2 surface.

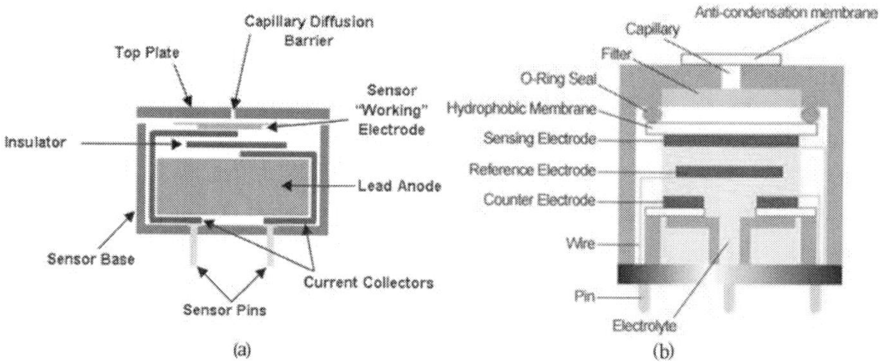

Figure 2. (a) Amperometric cell sensor and (b) potentiostatic cell sensor.

precise measurements at low gas concentrations and the calibration procedure becomes a lot simpler calibration. The diffusion control also allows the sensor manufacturer to tune the sensor to a particular target gas concentration range by simply changing the diffusion barrier. Finally, the calibration of electrochemical sensors could be more stable on time and decrease the operation costs due to less maintenance needs. The sensor sensitivity could be calculated based on the diffusion properties of the gas path into the sensor. However, attention has to be paid to the possible experimental errors in the measurement of diffusion properties [4].

The detailed manufacturing of these electrochemical sensors comprises the use of a membrane filter (MEMBRAPOR), which performs a preliminary cleaning of the target gas molecules from dust and water vapors; then the gas flow is directed through a capillary to a second filter and finally through a hydrophobic membrane, before reaching the sensing electrode surface. On the sensing electrode, the detected molecules are readily oxidized or reduced on the active catalytic sites from the surface, generating or taking electrons according to the behavior of the species to be detected (reducing or oxidizer gas) and generating an electric current. This approach is used to control mechanically the amount of gas molecules entering the sensor by diffusion through the capillaries. The simple principle of electrochemical detection is respected, and the sensing optimization can be done in line with the desired measurement range, by regulating the gas amount reaching the sensing area till the electrical signal value considered adequate is obtained.

The role of the catalytically active surface is expected to reach a high reactivity toward the target gas and inhibit the possible undesired interfering responses; therefore, the chemical composition, preparation method and activation of the sensitive solid are crucial in designing the sensing electrode.

An important advantage of the MEMBRAPOR gas sensor is that the signal obtained is connected rather to the target gas concentration than to its partial pressure. So, the sensor can be used at different altitudes (even underground) and the results are independent of the pressure when the device was calibrated.

One of the most common and widely used amperometric sensorsis the technical device installed currently on the exhaust pipe from automotives, named lambda probe. The role of the device is to measure the oxygen concentration after the engine. This value is related to the correct air-to-fuel ratio in the engine and determines the pollution level of the combustion. The air-to-fuel ratio can be regulated by the air admission in the engine and is controlled by the electronic control unit of the car. The proper air-to-fuel ratio (as near as possible to the stoichiometric ratio) is necessary to settle, in order to avoid both the nitrogen oxides formation from the air oxygen and nitrogen, on one hand, and for the complete burning of the fuel, on the other hand.

The working principle of the lambda probe is the measurement of the electromotive force (EMF) generated across a layer of ion solid electrolyte (usually, yttria-stabilized zirconia, YSZ) included in an electric circuit. The trivalent dopant from the framework of the tetravalent zirconium oxide generates oxygen vacancies, on which oxygen can be adsorbed as O^{2-} ion at temperatures above 300°C:

$$O_{2(gas)} + 4e^- \rightarrow 2O^{2-}{}_{(electrolyte)} \qquad (r-3)$$

The solid structure of the semiconductive material allows a high mobility of the O^{2-} ions (behaving as a solid electrolyte), so they migrate through the YSZ layer from one face to the opposite one, where they lose the captured electrons and desorbs as O_2 molecules.

$$2O^{2-}{}_{(electrolyte)} \rightarrow O_{2(gas)} + 4e^- \qquad (r-4)$$

This couple of reactions can be valorized in terms of sensor signal either by a potentiometric device or by the amperometric method.

The electromotive force (E) induced across the electrolyte can be expressed by the relation:

$$E = \frac{kT}{nq}\int t_{ion} d(\ln P_{O_2}) \qquad (1)$$

where k is the Bolzmann constant, T is the absolute temperature, n is the number of transferred electrons, q is the electron charge, and t_{ion} is the ionic transference number. The expression can be simplified for the equations above to the form:

$$E = \frac{RT}{4F} \ln \frac{p_1}{p_2} \tag{2}$$

where R is the universal gas constant, F is the Faraday number, and p_1 and p_2 are the partial pressures of oxygen on the two sides of the electrolyte. This relation is known as the Nernst equation.

When the oxygen concentration is measured by the potentiometric method, the electromotive voltage U associated to the oxygen ions circulation within the sensing layer generates is measured. The value of U is of course proportional with the difference between the partial pressure of oxygen on the two sides of the zirconia layer [5].

When the amperometric method is used, a current applied between the two sides of the sensing layer crossed by oxygen intensifies the circulation of the O^{2-} ions and the intensity due to their migration can be measured. Zirconium oxide doped with yttrium oxide is thus used as an "oxygen pump" transporting the gas from one face of the sensing layer to the other. The current intensity thus generated is a measure of the amount of oxygen transported through the layer [6].

The adequacy of a potentiometric oxygen sensor versus an amperometric one depends on the expected oxygen concentration. When the concentration range is wide (case of gasoline engines, where the oxygen ratio must be near the stoichiometric ratio), potentiometric sensors are preferred, despite their intrinsically low sensitivity. In the case of Diesel engines, in order to avoid soot formation, a lean-burn ratio is used; therefore, an amperometric mode is utilized for oxygen detection. This relies on the fact that a limiting current, proportional to the partial pressure of the gas, but independent of applied voltage, is reached when mass transfer across the electrolyte becomes limited by diffusive flux of gas toward the electrode, where the electrochemical reaction takes place, generating this current. The sensing principle is sketched in **Figure 3**.

Sensors based on this principle are now being applied in diesel-engine equipped cars operating under lean-burn conditions. A key challenge for such sensors is that the limiting current depends on the morphology and catalytic activity of the electrodes, which change in time, during operation.

This type of sensor has a good reproducibility, adapts easily to various design versions and works in a wide range of oxygen pressure values. Drawbacks and limitations include the high temperature required for a good working function, the difficulties connected to the sealing of the measuring cell and the interference of other gases possibly present in a real environment mixture. Another attention reason is that the concentration of partial pressure of oxygen in the exhaust gas from engines changes very much (several orders of magnitude). This type of sensor is usually included as the so-called lambda sensor to eventually detect the remaining oxygen in the exhaust gases from automotives. The information from the lambda probe is

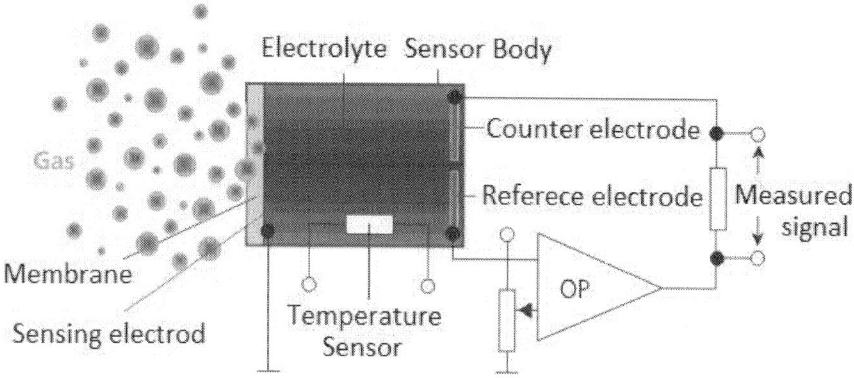

Figure 3. Sensor for automotive exhaust gases.

transmitted to the electronic control unit (ECU) of the automotive, which regulates the air-to-fuel ratio in the engine. Therefore, the sensor response must be as fast and sensitive as possible.

3. Influence of the sensing surface micro- or nanostructure

When the sensing layer acts as a catalyst that can promote a reaction between species from the gaseous environment of the sensor, the receptor will suffer changes in the layer charge, due to the electron transfer associated with the chemical reaction. These changes can be transformed by the transducer in a measurable signal. Usually, combustion reactions of several organic species or carbon monoxide generate high values of the thermal effect; therefore the reaction heat can also be measured and used to deduce the amount of combustible sensed species. In the combustion process, the oxygen reacts in adsorbed state.

Most sensors based on semiconductive oxides show improvement when doped with noble metals. Thus, a so-call "triple-phase boundary" (TPB) is formed where the gas phase meets both semiconductor oxide and metal. The interface between oxide and metal has different extents of incomplete orbitals or electrons which can be transferred to appropriate species. Thus, oxygen can transform on the active sites (*) at the interface in electrophilic O_2^-, O_2^{2-} or in nucleophilic O^{2-}. The sensing mechanism highly depends on the nature of the adsorbed oxygen [7, 8]. When a reducing species exists in the gas to be detected/measured, this species could react with the adsorbed oxygen directly from gas phase (Elley-Rideal mechanism) and form carbon dioxide. The electrons provided by O^{2-} existing as a mobile species in the solid electrolyte will be released and the solid conductivity increases. The following reactions can be written to support the above statements:

$$O_{2(gas)} + 2* \leftrightarrow 2O_{(ads)} \qquad (r-5)$$

$$O_{(ads)} + V + 2e^- \leftrightarrow O^{2-}{}_{(electrolyte)} + * \qquad (r-6)$$

$$CO_{(gas)} + O_{(ads)} \leftrightarrow CO_{2(gas)} \qquad (r-7)$$

$$CO_{2(gas)} + O^{2-}{}_{(electrolyte)} \leftrightarrow CO_{2(ads)} + V + 2e^- \qquad (r-8)$$

In a sensor for CO, since (r-7) is a cathodic reaction and (r-8) is an anodic one, the transducer transmits, in fact, the sum of signals due to oxygen adsorption and CO oxidation. In these cases, it is important to prepare the material in an appropriate manner so that the metal-oxide joining in the TPB improves and preserves the sensitivity and selectivity of the sensor.

This kind of practical approach can also be used in hydrogen and ammonia sensors, since these species can decompose on Pt, Pd or Au due to the spillover of hydrogen atoms to the metal oxide. This hypothesis is supported by Barbosa et al. [9], which prepared SnO_2 micro-disks made of tin oxide nanoparticles from the bulk oxide and carbon black, following a thermal treatment at 1135°C, then doped the resulted material with Ag or with Pd. Two distinctive mechanisms were proposed for the detection on the pure and doped oxide, since the ability of Pd to split in atoms of the hydrogen molecule involves its participation in adsorbed state. Thus, for pristine SnO_2, the proposed reactions are:

$$2H_{2(g)} + O^-{}_{2(SnO)} \rightarrow 2H_2O_{(g)} + e^-{}_{(SnO)} \qquad (r-9)$$

$$H_{2(g)} + O^-(SnO) \rightarrow H_2O_{(g)} + e^-{}_{(SnO)} \qquad (r-10)$$

The nature of ionic oxygen species from the solid surface depends on an important factor, on the temperature value; at low temperature, the O_2^- species predominates, while at higher values, O^- is the major species. Hydrogen as free gas can be easily oxidized by the adsorbed oxygen. The released electron decreases the resistance value of the sensing layer, so the R_{air}/R_{H2} ratio increases proportionally with the hydrogen concentration (**Figure 4**). Sensor response when SnO_2 was doped with Ag was insignificantly modified, while when doping was made with Pd, the sensitivity toward hydrogen increased a lot. Therefore, the authors propose another mechanism for the sensing process, based on the hydrogen splitting on Pd sites, followed by the oxidation on SnO sites and release of a higher number of electrons, suffering a spillover effect:

$$H_{2(g)} + Pd \rightarrow 2H_{(Pd)}(splitover) \qquad (r-11)$$

$$4H_{(Pd)} + O^-{}_{2(SnO)} \rightarrow 2H_2O_{(g)} + e^-{}_{(SnO)}(spillover) \qquad (r-12)$$

$$2H_1 + O^-{}_{(SnO)} \rightarrow H_2O_{(g)} + e^-{}_{(SnO)}(spillover) \qquad (r-13)$$

The sensor response is a lot higher in terms of R_{air}/R_{H2} ratio on Pd-doped SnO_2 and is highly favored by the temperature increase up to 300°C; this is in fact the upper value allowing the hydrogen adsorption.

The detection of nitrogen oxide is of crucial importance in the environment monitoring. Nitrogen oxides, NO and NO_2, denoted generally as NO_x, are formed during the fuel burning in stationary sources (energy plants) and in the automotives' engines, from the elements of the air, in the harsh conditions (high temperatures, especially in case of Diesel engines). NO has

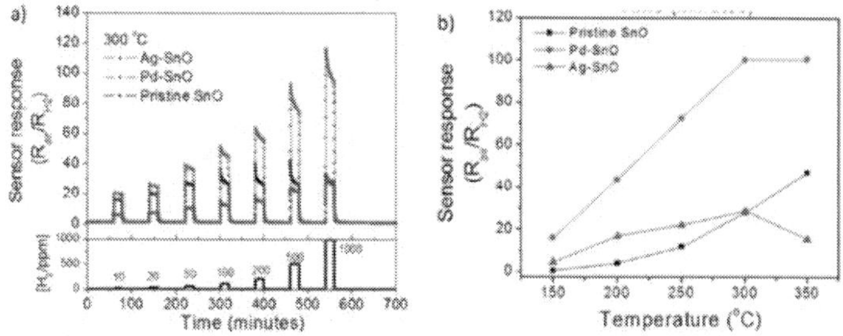

Figure 4. Hydrogen sensing on SnO_2 and Pd-doped SnO_2 disks of nanoparticles. (a) Sensor response at 300°C at 1000 ppm hydrogen. (b) Temperature dependence of the sensor response.

character of reducing agent and NO_2 acts rather as an oxidant. The NO detection mechanism supposes the participation of O^{2-} by the reaction:

$$NO + O^{2-} \rightarrow NO_2 + 2\,e^- \quad (r-14)$$

while NO_2 decomposes to NO, it can be detected in the same manner:

$$NO_2 + 2\,e^- \rightarrow NO + O^{2-} \quad (r-15)$$

If the sensing layer contains basic sites (In^{2+} and Sn^{2+} ions, for instance), NO_2 is adsorbed as an acid anhydride and decomposed to NO and nitrogen:

$$NO_{2(gas)} + Me^{2+} \rightarrow (Me^{2+} - NO_{2(ads)}) \rightarrow (Me^{3+} - O_{ads}{}^-) + NO_{(gas)} \quad (r-16)$$

$$NO_{(gas)} + Me^{2+} \rightarrow (Me^{2+} - NO_{(ads)}) \rightarrow (Me^{3+} - O_{ads}{}^-) + ½N_{2(gas)} \quad (r-17)$$

The reactions occur better on stronger basic sites, because of the competition between the two oxygen species $O_{2(ads)}{}^-$ and $O_{(ads)}{}^-$ and NO_2. On less basic sites, NO_2 can be oxidized to $NO_3{}^-$:

$$NO_{2(gas)} + O_{(ads)}{}^- \rightarrow NO^-{}_{3(ads)} \quad (r-18)$$

Indium oxide has a somehow basic character, so its sites will act rather as adsorption sites for NO_2. On the contrary, on SnO_2 the reaction between NO_2 and $O_{(ads)}{}^-$ is by far preferred. The prevalence of these two reactions depends on the temperature value. The role of the basic sites is more significant at low temperatures. The mixed indium an tin oxide (ITO) brings the advantages of both components: conductivity (due to the presence of the two metallic species) and high defect density, bringing active adsorption and reaction sites.

Saboor et al. [10] prepared nanoparticles of SnO_2 to be used as sensing material and stimulated the chemical interaction between the solid and the sensed species (NO_2) by UV irradiation. The tin oxide samples were prepared from $SnCl_2$ or $SnCl_4$, by precipitation in hydrothermal conditions at 190 and 240°C, using ethanol and/or NaOH solution to initiate precipitation, or by the action of microwaves on a mixture of $SnCl_4$, NaOH, and ethanol. The samples were used either as-synthesized or after calcination at 400 or 500°C. The resulted powders were screen-printed on interdigitated electrodes deposited on alumina substrates. The sensing properties were tested by measuring the electrical resistance of the SnO_2 material deposited between the metallic contacts in the presence of 5 ppm NO_2, in comparison with pure air, and the response was expressed as the resistance ratio R_{gas}/R_{air}. For a sensing material, it is also important to investigate the recovery time after sensing, i.e., the time needed for the baseline to return to the initial value, after cutting off the contact with the sensed species; generally, the recovery time is the time value needed to decrease 80–90% of the response value.

The results in the sensing procedure indicate that pristine SnO_2 gave almost no signal without UV light, while the signal magnitude depended strongly on the irradiation intensity. This effect must be explained in connection with the sensing surface microstructure (which is correlated with the value of the specific surface area and free pore volume), more precisely, by the presence of a high number of grain boundaries and by the role of UV light in the charge carriers generation during the irradiation of a semiconductive material. The experimental results indicate that the semiconductive properties of tin oxide depend on the sample structure and morphology, i.e., on the preparative method. The SnO_2 surface is covered with adsorbed oxygen species O_2^-, O^- and O^{2-}, the first being predominant at temperatures below 100°C. Under UV irradiation, on one hand, the oxygen adsorption is favored, and on the other hand, the electron-hole pairs' formation and separation, characteristic to the semiconductive oxide photo-generated charge separation, is stimulated by the built-in electric field near the surface of SnO_2 nanoparticles. This increase in the number of charge carriers is reflected because of a drastic decrease in the electrical resistance in air. When NO_2-containing air flows on the sensor area, the molecules pick up electrons from the surface and therefore, decrease the number of free carriers, i.e., decrease the electrical resistance value. The following reactions occurring on the surface can explain the observed transient responses, as well as the effect of irradiation on the sensing signal magnitude (**Figure 5**).

$$NO_2(g) + e^-(h\upsilon) \rightarrow NO_2^-{}_{(ads)} \quad \text{(r–19)}$$

$$NO_{2(g)} + O^-{}_{2(ads)} + 2e^- \rightarrow NO^-{}_{2(ads)} + 2O^-{}_{(ads)} \quad \text{(r – 20)}$$

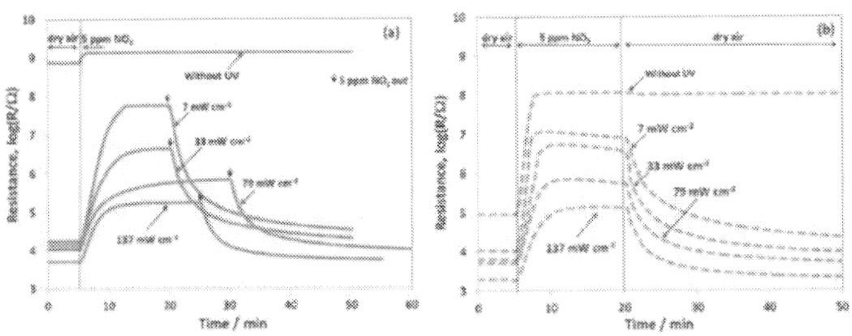

Figure 5. Values of the electrical resistance in time on 5 ppm NO_2 exposure under UV irradiation, on as-synthesized (a) and calcined (b) SnO_2 prepared hydrothermally from $SnCl_4$.

The results indicate that the sensing is favored only when the sensor area is irradiated by low intensity UV light, in terms of both sensing performance and recovery time of the sensor. The explanation lies in the moderate activation of the surface by UV light, in terms of formation of active species of oxygen. The higher UV light intensity may contribute to the desorption of the chemisorbed oxygen.

4. Main types of electrochemical sensors used in practical applications

4.1. Sensors for toxic gases

Electrochemical sensors destined for toxic gases detection and measuring work as micro fuel cells, are designed as maintenance-free and stable devices for long periods. A simple sensor with this purpose consists of two electrodes, a sensing and a counter one, allowing placing a thin layer of electrolyte between them (**Figure 6**). The electrodes are accommodated in a plastic box with a capillary opening to allow the gas entry to the sensing electrode. The electrodes are connected through two pins going to the external interface and thus allow the sensor connection in a resistor circuit for measuring the voltage drop resulting from any current flow.

The molecules of the target gas diffuse into the sensor and are oxidized or reduced on the sensing electrode. On the other hand, on the electrode takes place the corresponding inverse reaction, and consequently, a current is generated in the external circuit. The amount of gas entering in the sensor depends on the capillary diffusion barrier. The current generated in the circuit is proportional to its concentration, so that the sensor gives a direct measure of the interesting species of toxic gas; the response depends rather on the volume or concentration of gas than on its partial pressure.

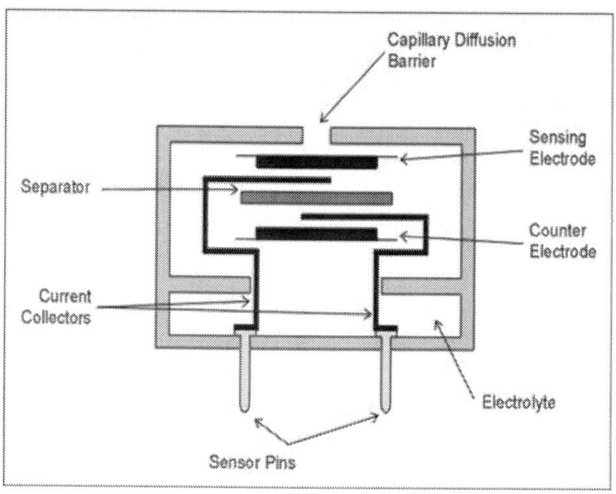

Figure 6. Toxic gas sensor.

The key feature of this sensor design is thus the gas diffusion barrier, whose role is to control/limit the amount of gas reaching the sensing electrode. On the electrode, all target gas admitted inside the cell will practically react. Common commercial products for measuring various toxic gases are the CiTiceL sensors. The reactions occurring at the electrodes in sensor exposed to a gas flow containing carbon monoxide are:

$$\text{Sensing}: CO + H_2O \rightarrow CO_2 + 2H^+ + 2e^- \qquad (r-21)$$

$$\text{Counter}: 1/2\, O_2 + 2\, H^+ + 2e^- \rightarrow H_2O \qquad (r-22)$$

$$\text{The overall reaction is}: CO + 1/2\, O_2 \rightarrow CO_2 \qquad (r-23)$$

The detection principle for other toxic gases which can be oxidized electrochemically is similar. The reactions (r-21)–(r-23) above indicate that the oxygen is necessary on the counter electrode to generate the current. Usually, the sample stream readily contains oxygen. The sensor exposure to an anaerobic sample gas may result in signal drift; therefore, the toxic gas sensors should never be immersed in an anaerobic gas mixture. In some applications, the sensors could be frequently exposed to high concentrations of analytes; in this case, it is necessary to ensure an additional oxygen source near the counter electrode. In the gas sensing procedure, it is important to make sure that the rate-limiting step is the gas diffusion through the capillary, while all other stages are a lot faster. A way to achieve this goal is to use electrode material as solid with high catalytic activity for the electrochemical reactions based on the sensor.

4.2. Three- and four-electrode sensors

The simplicity of design of the two-electrode sensors, however, brings a major drawback: the polarization of the counter electrode, which limits the measuring range. This effect could be eliminated by using a third electrode, named reference electrode, having a stable potential. It is held at a fixed potential relative to the reference electrode (no current is drawn from it), so both keep a constant potential. The counter electrode can still polarize but will not affect the sensing electrode. The three-electrode sensors are generally used in the electrochemical sensors for detecting toxic gases.

In some applications, however, the presence of the third sensor brings changes that compromise the sensor behavior, as the cross-interfering gases or zero-offset changes with temperature. In this case, the introduction of a fourth "auxiliary" sensor allows the sensor performance to be kept in order to allow the simultaneous measurement of two gases.

4.3. Overcoming cross interference

The fourth auxiliary electrode can be useful in preventing the cross-sensitivity to other gases. For example, the sensing reactions involved in carbon monoxide sensors are similar in terms of electrical changes of the sensing area to those found in the hydrogen gas sensing. Therefore, if hydrogen is present in a gas mixture containing carbon monoxide, the response of hydrogen damages accurate measurement of CO. In this case, the forth-auxiliary electrode has the role to measure the hydrogen, while CO and some of the H_2 reacts on the sensing electrode. The signal from the forth electrode is used as a compensating signal, which is subtracted from the signal of the sensing electrode using a microprocessor with appropriate software.

4.4. Overcoming temperature effects

The temperature effect is due to the important increase of the signal when temperature changes. A 10°C rise of temperature almost doubles the baseline signal of most electrochemical sensors. The effects are significant especially when measuring low concentrations of gases, such as ozone and carbon monoxide. Since the signals from both sensing electrode and auxiliary electrode give similar responses to changes in temperature, the signal from the auxiliary electrode can simply be subtracted from that of sensing electrode to compensate the temperature effect.

4.5. Dual/multiple gas sensor

The multiple sensor modules find applications in the monitoring of the urban atmosphere, the environment associated with energy plants or other industrial units, the unorganized emissions, the emergency monitoring, the land monitoring using portable gas detector for various gas detection equipment as well as smart home appliance.

An example of sensor widely applied in gas monitoring is the **4COSH CiTiceL** four-electrode sensor, manufactured by City Technology company from the Great Britain [11]. It allows measuring CO and H_2S simultaneously with one sensor with a portable instrument. The sensor has wide ranges of measurable gas concentration values: 0–500 ppm CO and 0–200 ppm H_2S, at resolution values of ± 1 ppm for Co and ± 0.5 ppm for H_2S in a relative humidity range of 15–90%. It is important to mention that the declared signal loss is 5% per year. The weight is only 5 g, and the gauge size is only 2×1.6 cm.

The sensor includes two sensing electrodes, one for CO and the other for H_2S. On the first sensing electrode, hydrogen sulfide is oxidized completely and CO diffuses through the sensor being oxidized on the second electrode. This sensor obtains two separate signals, one for each gas, but allows two gases to be measured with one sensor.

ZE10 (**Figure** 7) is a multispecies, high-performance electrochemical device for gas sensing, able to detect CO, SO_2, NO_2 and O_3, with good selectivity and stability, produced by Winsen Company (China) [12]. It has a built-in temperature sensor for thermal compensation. It combines the electrochemical detection principle with sophisticated electrical circuit design, aiming to meet different detection requirements from customers. The detection resolution is up to 10 ppb, the response time 30 s, the humidity range 15–90%. In **Figure 8**, the electrical diagram for a potentiostatic sensor is presented.

In **Table 1**, the technical data about the sulfur dioxide and nitrogen dioxide sensors are exemplified.

The Figaro company (USA) is producing various types of gas sensors: metal oxide sensors (MOS), catalytic sensors, and electrochemical sensors. Figaro electrochemical sensors are amperometric fuel cells with two electrodes, the working (sensing) electrode and the counter

Figure 7. Multielectrode electrochemical sensors for detection of H_2S and CO.

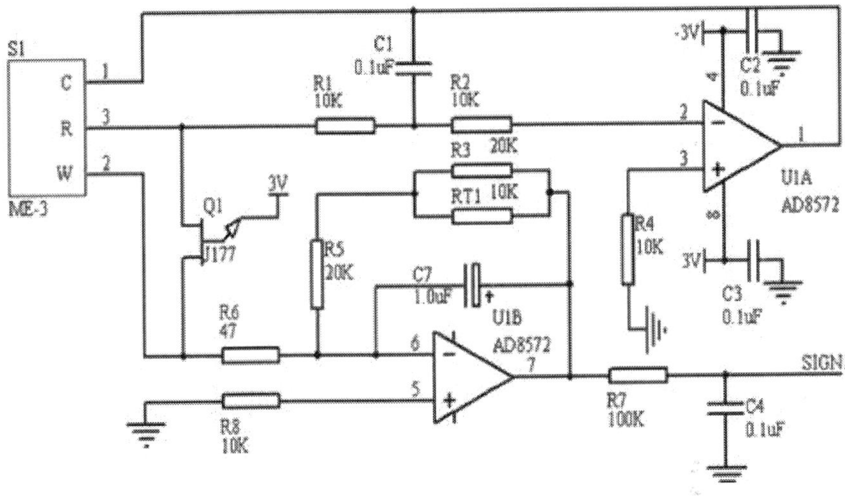

Figure 8. The electrical diagram for the ME-3 potentiostatic sensor (3 electrodes: C, R, and W).

electrode, having placed in between a layer of ion conductor (**Figure 9**). When a reducing gas (CO) reaches the working electrode, its oxidation will occur on it, with the participation of a water molecules from the surrounding air (r-24).

$$CO + H_2O \rightarrow CO_2 + 2H^+ + 2e^- \qquad (r-24)$$

Since the working electrode is connected to the counter electrode, the protons (H^+) formed by reaction (r-24) will flow toward the counter electrode through the ion conductor and the electrons will move to the counter electrode through the external wiring. The protons and electrons will then react with oxygen from the environmental air (r-25)

$$1/2O_2 + 2H^+ + 2e^- \rightarrow H_2O \qquad (r-25)$$

The overall reaction is thus:

Parameter	SO_2 sensor, sulfur dioxide	NO_2 sensor, nitrogen dioxide
Measurement range	0–20 ppm	0–20 ppm
Max detecting concentration	150 ppm	150 ppm
Sensitivity	(0.55 ± 0.15) μA/ppm	(0.78 ± 0.42) μA/ppm
Resolution ratio	0.1 ppm	0.1 ppm
Response time (T90)	<30 s	<25 s
Bias voltage	0 mV	0 mV
Load resistance (recommend)	10 Ω	10 Ω
Repeatability	<2% output value	<2% output value
Stability (/month)	<2%	<2%
Output linearity	Linear	Linear
Zero drift (−20–40°C)	≤0.2 ppm	0.2 ppm
Storage temperature	−20–50°C	−20–50°C
Storage humidity	15–90% RH	15–90% RH
Pressure range (kPa)	90–110	90–110
Anticipated using life	2 years (in air)	2 years (in air)

Table 1. Technical parameters for some electrochemical gas sensors produced by Winsensor.

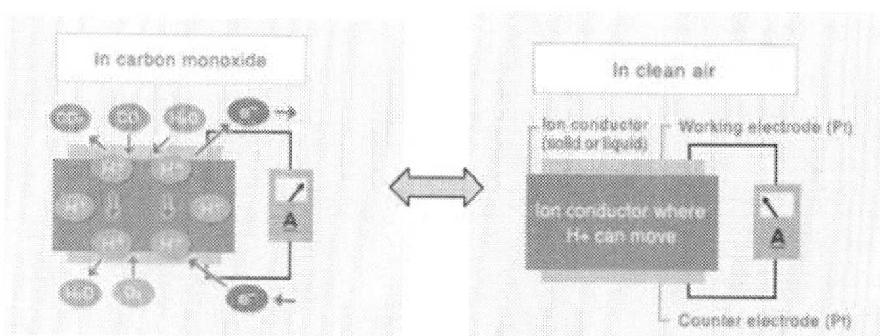

Figure 9. Changes and moving of the charge carriers in a Figaro electrochemical gas sensor exposed to CO-containing gas and clean air.

$$CO + 1/2O_2 \rightarrow CO_2 \qquad (r-26)$$

The simplicity, robustness and low piece of the Figaro sensor helped it to be present in most domestic and industrial applications, especially for the detection of organic compounds associated with the fuel gas leaking: residential gas detection, leak checkers.

4.6. Electrochemical sensors working at high temperatures

A very important practical application of amperometric method in gas sensing is found in the production of oxygen sensors, able to work at high temperature, to monitor the composition of exhaust gas from automotives. The precise monitoring of the oxygen in the exhaust gas allows

Chapter 5

Experimental Analysis of Modified CNTs-Based Gas Sensor

Ju Tang, Xiaoxing Zhang, Song Xiao and
Yingang Gui

Additional information is available at the end of the chapter

http://dx.doi.org/10.5772/intechopen.68590

Abstract

As a significant equipment in power system, the operation condition of transformers directly determines the safety of power system. Therefore, it has been an indispensable measure to detect and analyze the dissolved gases in transformers, aiming to estimate the early potential faults in oil-insulated transformers. In this chapter, the adsorption processes between modified carbon nanotubes (CNTs) (CNTs-OH, Ni-CNTs) and dissolved gases in transformers oil including C_2H_2, C_2H_4, C_2H_6, CH_4, CO, and H_2 have been simulated based on the first principle theory. Meanwhile, the density of states (DOS), adsorption energy, charge transfer amount, and adsorption distance of adsorption process between CNTs and dissolved gases were calculated. Moreover, two kinds of sensors, mixed acid-modified CNTs and $NiCl_2$-modified CNTs, are prepared to conduct the dissolved gases response experiment. Then, the gas response mechanisms were investigated. Finally, the results between response experiment and theoretical calculation were compared, reflecting a good coherence with each other. The CNTs gas sensors possess a relatively high sensitivity and fine linearity, and could be employed in dissolved gas analysis equipment in transformer.

Keywords: oil-insulated transformer, modified CNTs, dissolved gases, DFT method, gas-sensing experiment

1. Introduction

1.1. The significance of dissolved gases detection in oil-insulated transformer

The stability of electrical equipment is the key factor for the running safety and economy of electrical power system. Along with the extra-high-voltage grid construction, power transmission capacity becomes larger, coverage area becomes wider, and the national power grids at all levels are closely linked to each other, so the harm of grid accidents would be more serious [1]. Large power transformer, as the key equipment of power system, plays influential role to ensure the safe operation of power system. Real-time detection of insulation state of transformer, accurately predicting the fault and avoiding possible trouble are important measures to ensure the safe operation of the electrical grid [2], to improve equipment utilization and reduce costs of equipment maintenance, which are also key technical issues of constructing the strong and intelligent electrical grid. A lot of study and practice has proved that the main reason for the transformer accident is the deterioration of its insulation performante. With the development of electronic, computer, sensor and information processing technology recently, detection ways for insulation state of large-scale power transformer have been rapidly developing. For example, dissolved gas analysis (DGA) [3], partial discharge (PD) [4], winding coefficient of dielectric loss measurement, winding insulation resistance measurements, winding deformation and winding hot spot temperature monitoring the micro water insulation monitoring, etc. [5], these ways can help people to get the insulation state of transformer from different aspects.

Dissolved gas-in-oil analysis (Oil-DGA) is the most convenient and effective method of judging the early potential fault of oil-immersed power transformers at present [6–8], and the method is the most extensive one in the real application, which has become an indispensable approach to judge the internal fault of oil-filled electrical equipment and oversee the safe operation of equipment [9]. Research shows that, among main obstacle and defects of transformers discovered by experiment examinations, faults found by dissolved gas analysis of test standard always take up the highest percentage, 60.1% in 2004 and 68.5% in 2005 [10]. The international electro technical commission develops standard IEC60567 "Oil-filled electrical equipment—Sampling of gases and analysis of free and dissolved gases—Guidance" and IEC60599 "Mineral oil-impregnated electrical equipment in service—Guide to the interpretation of dissolved and free gases analysis." Gas chromatography is widely applied to quantitative analysis of various gases dissolved in transformer oil content. For the past few years, it has become the new trend to develop the pint-sized gas-detecting device by using the gas-sensing technology, which is aimed at achieving on-line monitoring to dissolved gas in transformer oil and grasping the operational state of equipment at any time [11, 12]. On-line monitoring of dissolved gas analysis could help reduce the unavailability of equipment in the long run-time [13], thus it can improve the economic benefit, optimize cycle and content of maintenance job to decrease maintenance fee and improve control of power system and the reliability of monitoring performance in overload operation [14].

The high-sensitivity gas sensors are used to detect dissolved gas in transformer oil [15], structure of test system is simple and it is easy to implement. In recent years, a breakthrough has been making in the sensor technology [16, 17]. Especially, the development of nanotechnology has been providing the new material and processing method [18], in which the carbon nanotubes (CNTs) gas sensor has become the new research focus [19]. CNTs have abundant pore structure, large specific surface area, and a strong ability of adsorption and desorption for chemical composition of the gas phase, these properties make CNTs, as gas sensor, incomparable in conventional sensors of the detection sensitivity and miniaturization [20]. At present, this technology obtains rapid development in the biological, chemical, machinery, aviation, military, and other aspects.

As the pressure of global resources and environment is increasing, the society demand for environmental protection, energy conservation, and emissions reduction and sustainable development is increasing day by day. Along with sustainable development of social economy, the rapid growth of electricity demand, carrying out the energy conservation and emissions reduction and construction of "resource saving and environment-friendly society," has become a very urgent task. The Electric Power Research Institute (EPRI) put forward the concept of smart grid in 2000; they think this is development tendency in the future power grid and a way to solve the problem of the grid in twenty-first century. So-called smart grid is the advanced sensor measurement technology, information technology, communication technology, computer technology, automatic control technology, and the original transport and distribution infrastructure highly integrated to form a new type of power grid. The observability based on advanced measurement, sensing technology, and real-time analysis based on comprehensive analysis of decision-making reflects mainly intelligence in grid [21, 22], and it is one of the hot topics in the study of the current smart grid.

In conclusion, new sensors researched are used to detect equipment on real-time and accurately predict failure, do nip in the bud. These are important measures to ensure the safety of the grid in production, improve equipment utilization, and reduce the equipment maintenance cost, also, is the key technology of construction for unified strong smart grid. Based on the study of the existing transformer oil gas detection and analysis method, this chapter studies deeply application of nanometer gas-sensitive sensor technology for transformer oil gas detection and analysis, develops a new CNTs gas sensor and, tests its gas-sensing property, master the basic law, put forward an algorithm of dissolved gas analysis in oil-based dynamic tunnel fuzzy c-means to rich analysis method of transformer fault characteristics. This research topic from the urgent demand in reality not only has important academic value but also has significant economic and social benefits and broad application prospects.

1.2. The methods for dissolved gases detection in oil-insulated transformer

At present, most of high-voltage and large-capacity power transformers used oil filled; this transformer used composite structure including insulating oil and insulation paper (plate) and has the very high-electric strength. During long-running process, insulating oil and other insulating materials of transformer under the effect of electricity and heat will gradually age and decompose, resulting in the production of gases such as low molecular hydrocarbons, CO, and CO_2 that get dissolved in insulating oil [23, 24]. When the transformer is in normal operation condition, the outside factors such as electricity, heat, and mechanical stress cannot break chemical bond of insulating oil and insulation paper (plate). Insulating materials only produce little gas when they are normal aging. But the discharge and overheating fault occur in the equipment, deterioration process of insulating materials is greatly accelerated to accelerate rate of biogas production of the above gas [25]. The study demonstrated composition and content of these gas has close relation with property of fault, these gases are called as characteristic gases, as shown in **Table 1**.

It has a great significance for indication of the transformer early fault to monitor change of characteristics of the gas composition and content in insulating oil. Electromagnetic interference shows no influence on dissolved gases analysis, and dissolved gases analysis shows high data reliability.

Fault type	Insulation medium	Main components	Minor components
Overheating	Oil	CH_4, C_2H_4	C_2H_6, H_2
	Transformer oil paper insulation	CH_4, C_2H_4, CO, CO_2	C_2H_6, H_2
Arc discharge	Oil	C_2H_2, H_2	C_2H_6, CH_4, C_2H_4
	Transformer oil insulation paper	C_2H_2, H_2, CO, CO_2	C_2H_6, CH_4, C_2H_4
Partial discharge	Transformer oil and insulation paper	CH_4, CO, H_2	C_2H_6, CO_2
Spark discharge	Transformer oil	C_2H_2, H_2	/
Bubble discharge	Watered oil	H_2	/

Table 1. Characteristic gases of several styles of faults.

The technology develops maturely and accumulates considerable experience from qualitative to quantitative analysis to form relative regulation, such as International Electrotechnical Commission standard IEC60567 and IEC60599. Since the 1970s, transformer fault diagnosis based on characteristic gases dissolved in transformer oil has widely spread all over world and gradually applied in the field.

According to "the guidelines for the dissolved gas in transformer oil analysis, and judgment," detection of gas content first needs gas separation from oil (i.e., the degassing). The gas separated from oil is a mixture of various gas components. The general gas content detection method has sensitive for all kinds of gases or a gas. To detect a single gas, we usually use gas chromatography to separate various gas components [26]. The advantage of gas chromatographic method is quantitative analysis for a variety of gas content dissolved in oil; meanwhile, there are characteristics in it, such as many test links, complex operation, high technical requirements, and long test cycle. Therefore, this method is usually used to regular checking for main equipment (e.g., once half a year), professional workers conduct it in laboratory. However, in two regular interval periods, the internal condition changes of transformer cannot be found in a timely manner. Application of micro gas-sensor technology is developed to miniaturized gas detection device, which can monitor on-line dissolved gas in transformer oil to master the running status of equipment whenever. When an alarm occurs in on-line monitor, we could use method such as chromatographic analysis to secondary diagnosis [27].

The on-line monitor of dissolved gas in transformer oil still based on the categories and quantities of dissolved gas which can be seen as fault characteristic quantities. The difference is that real-time on-line monitoring of oil chromatography and intelligent fault diagnosis can be realized using this technology. This can not only gain the running state of the transformer timely, based on which latent faults can be detected and tracked, but also diagnose fault automatically due to the expert system so as for operating crew, fault can be handled rapidly. Using on-line monitoring device can improve the management level of substation operation and lay a foundation of transition from preventive maintenance system to predictive maintenance system [28].

Choosing different sensors and cooperating to use different ways to take air and use diagnostic devices due to different test object can make up a variety of on-line or portable monitoring device to detect the dissolved gases in transformer oil [29, 30]. The method of detecting the dissolved gases in transformer oil can be divided into the following categories:

1. According to the categories of test object

a. Measuring the total content of combustible gases

The amount of combustible gases refers to the total quantities of H_2, CO, and all kinds of gaseous hydrocarbons. These kind of devices represented by TCG detection device of Japan's Mitsubishi electric power company can only give the amount of combustible gas but cannot measure the content of one component.

b. Measuring the content of single component: H_2

When overheating or partial discharge occurred in the equipment, hydrogen would appear. Fuel cell sensor such as HYDRAN produced by SYPROTEC Company in Canada can acquire signals. This device is suitable for the preliminary diagnosis for fault on the spot based on its simple structure, but chromatographic analysis must be applied to further determine the fault.

c. Measuring the content of composite gas components

With the development of on-line monitoring technology, on-line chromatographic detection devices have been invented for measuring full-component gases. The gas transformer oil on-line gas monitoring equipment of AVO Company in the USA can measure the contents of up to eight kinds of gases. DRMCC transformer on-line monitoring system can monitor the working status of transformer continuously, timely, and systematically. The main monitoring objects include dissolved substances such as hydrogen, water and wind temperature, position of tap, etc. The CONEDISON Company analyzed and measured the contents of CH_4, C_2H_4, CO, CO_2, and C_2H_6 using infrared spectroscopy method and measured the content of H_2 with an oxide electrochemical sensor. The on-line transformer fault prediction system developed by Chongqing university can measure the concentration of H_2, CO, CH_4, C_2H_4, C_2H_2, and C_2H_6 timely and can availably predict the concentrations of the dissolved gases in transformer oil and diagnose insulation condition of transformer in future with the method of gray clustering, paste pattern multi-level clustering, and kernel-based possibilistic clustering. Due to the limitation of sensing technology, the current on-line monitoring devices are not satisfactory in reliability and sensitivity, but on-line monitoring is the development direction of analysis technology of dissolved gases in transformer oil.

2. According to the categories of methods of extracting gas

a. Polymer separation membrane permeability method

Polymer membranes of organic synthesis have different degrees of permeability and can be used in industrial gas separation and purification process. In the mid-1960s, use of polyester hollow fiber membrane to recover hydrogen by Du Pont Company is one of the earliest attempts to use membrane to separate gases. The gases in oil follow Henry's law and go through the membrane into the air chamber so the gas concentration in the chamber and the dissolved gas concentration in the oil are balanced. The transmittance of the film is as high as possible to minimize the detection cycle, so that timely detection, timely alarm can be realized. Kurz first produced the polymer membrane and made its use for separation of transformer oil and gas, then the polyimide, polyhexafluoroethylene, polytetrafluoroethylene, and other polymer membranes were studied. China Electric Power Research Institute, Chongqing University, and other research institutions made repeated tests on the permeability of a variety of membranes, the results showed that Polytetrafluoroethylene (PTFE) film not only had good air permeability but also had good mechanical properties and resistance to oil, high temperature, and many other advantages. Therefore, it is often used as breathable film on the detectors of dissolved gas [31].

b. Method of vacuum pumping to extract gas

According to the principle of vacuum degassing, vacuum pump or bellows vacuum is used to extract dissolved gas in oil to achieve on-line monitoring of dissolved gas in transformer oil.

c. Other methods of extracting gas

Use different methods of blowing gas to replace oil-soluble gas so that the concentration of one kind of gas on surface and the concentration of that in oil gradually reach equilibrium and then analyze the gas on surface using a detector. Common methods contain carrier gas elution, air circulation, colorimetric pool method, etc.

3. According to the sensors that collect the signals

The detector that is used to detect separated or unseparated gases and has responses to a sample or samples is an important component of a detection device. It can be divided into palladium gate Field Effect Transistor (FET), semiconductor sensor, catalytic combustion sensor, combustion cell sensor, and other types of sensors [32].

The results show that combining two or more sensors, using modern computer technology, and developing the corresponding data processing software, we can measure the content of two kinds of fault gases and can achieve significantly better detection performance than one conventional sensor with a single monitoring device [33]. The development of gas-sensor array technology includes two aspects, one is to develop integrated micro gas-sensor array using micro-manufacturing, micro-machining technology; the other is to improve the accuracy of a single gas identification and realize quantitative analysis mixed gas using multi-sensor information fusion technology. Whether in the integration of sensor array, or in the analysis theory and technology of sensor array, it has become the hot spots of current sensor researches.

Using high-sensitivity gas sensor to detect dissolved gases in oil is easy to realize due to its simple gas line. But the existing gas sensor's detection sensitivity and reliability have not yet reached the level of sensitivity and reliability of off-line detection, so there are a lot of work to do. With the continuous development of detection technology, a variety of new sensors continues to come out, such as photoionization detector using energy of photons to ionize each type of gas, methane gas sensor, CO gas sensor based on vibration at room temperature, CO_2 gas sensor using solid battery, Pt doped SnO_2 separation membrane gas sensor by impregnation, CNTs sensor, etc. These new sensors create a good prospect of developing for the detection of high-sensitivity transformer oil-dissolved gas using on-line monitoring device. However, due to the harsh natural environment of on-line monitoring, complex strong electromagnetic interference, a large number of studies is also needed in selecting these sensors for gas detector [34].

2. Introduction of CNTs gas sensor

2.1. Introduction of gas-sensor technology and current situation

Sensor is a type of device that can transfer physical or chemical parameters into available electric information. Sensor can be defined as "device or apparatus which can sense the specified measurement and transfer it into available output signals by some laws" [35].

Gas sensors, namely gas-sensitive devices, are a type of device or apparatus that can sense specific gases and their concentrations in the environment. Information about the species and concentrations of gases can be transferred into electric signals for detection, monitoring, analysis, and alarm. Gas sensor is an important branch of sensor technology. Since the research on gas sensors started in the 1930s, it has passed more than half a century and there are several hundred kinds of sensors, which have been utilized in many aspects of human life, including national defense and military, industrial and agricultural production, energy and resource exploit, medicine, environment protection, disaster prediction, transport, etc. [36].

The fundamental characters of gas sensors include: sensitivity, selection, stability and resistant ability to corrosion, etc. Those characters are ensured by the selection of materials. Based on the characteristics of aimed gases, the environment conditions, detection requirements, and proper materials can be chosen or prepared for best gas-sensing properties of gas sensors. According to gas-sensing materials and gas-sensing response, gas sensors can be roughly divided into electric, optical, electrochemical, and other types, shown in **Figure 1**. The advantages and disadvantages of different gas sensors are shown in **Table 2**.

Microelectromechanical technology (MEMT) is the main manufacturing technology of gas sensors. MEMT is a type of new technology based on microelectronic technique and micromachining technology, including bulk micro-machining technology, surface micro-machining technology, and Lithographie, Galvanoformung and Abformung (LIGA) technology based on X-ray. Bulk micro-machining technology mainly aims at single silicon crystal, of which the key technique is corrosion and wire bonding technology with processing thickness of dozens to hundreds micron; surface micro-machining technology is based on semiconductor technique like oxidation, spread, photoetching, thin-film deposition, and other techniques, with thickness of several micron; LIGA technology adopts conventional X-ray for procession, with thickness of several to dozens micron. In those years, new processing techniques such as nanotechnology have provided more choices for sensors manufacturing technology and the development of processing techniques also motivates the breakthrough of sensor technology.

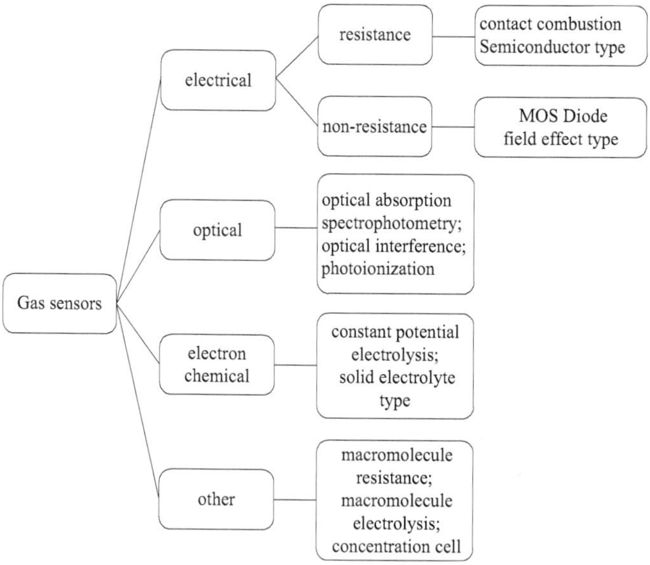

Figure 1. Classification chart of gas sensors.

Detectors	Advantages	Disadvantages
Field-effect tube detector	Only for H_2 detection, no interference by other gases	Not long lifespan, severe zero drift, false alarm
Catalysis incendiary detector [37]	Low cost, long lifespan, low effect by temperature and humidity, high-speed response, widely used for H_2, CH_4 detection	Not suitable for other gases, low gas selectivity
Semiconductor detector [38, 39]	By far the most widely used sensors, high response value, high response speed, good stability	Low gas selectivity
Combustion cell detector	Often used for H_2 detection, high detection precision, good repeat response capability	Limited lifespan, high cost and detection error
Infrared absorption detector	Often used for CO_2 detection, no sample separation	For useable for other gases
Optical gas sensor [40–42]	Electromagnetic insulation properties, high response speed, high response value, long lifespan, good stability	Complex detection system, high cost
Electrochemical gas sensor [43–45]	High response value, good gas selectivity	Easy influence from outside environment
High polymer gas sensor [46]	High response value to specific gases, good gas selectivity, important in food production	Only works under common temperature

Table 2. Comparison of the use features of common detectors.

2.2. Development status of CNTs-based gas sensors

Nowadays, with the development of industry manufacturing, environment detection as well as nanotechnology, nanogas-sensing technology has become the research focus of sensing technology. The development of nanotechnology provides not only excellent gas-sensing materials such as nanoparticles, nanowire, and nanoplane, but also new preparation and processing technology like scanning tunneling microscopy (STM) which enables researchers to observe the atoms and handle these atoms using probe. Therefore, this technology has developed a lot in biology, chemistry, machine, aviation, etc. Another important aspect of nanogas-sensing technology is CNTs gas sensors.

Since Iijima [47] found CNTs in 1991, both physical and chemical properties have been investigated widely. CNTs can be seen as tubes with nanosizes by graphite flake rolling. Hexagonal structural carbon atoms constitute several to 10 layers of tubes, with the distance between interfacing layers of 0.34 nm. There are single-wall CNTs (SWNT) and multi-wall CNTs (MWNT) according to the layers of CNTs as shown in **Figure 2**. The external diameter of CNTs is about several to dozens of nanometers and the length is about micron, much longer than the diameter.

CNTs possess abundant pore structure, large specific surface area, and excellent adsorption and desorption ability to gas molecules. Due to the interaction of adsorbed gas molecules with CNTs, the fermi level of it will change, thus leading to large macroscopic change of its resistance, which provides a way to detect the gases by the measurement of resistance. Those properties enable CNTs huge advantages as gas sensors. Firstly, large interaction surface provides gas adsorption sites with gas-sensing response enhanced largely. Secondly, working temperature can be lowered largely. Thirdly, gas sensors can also be controlled at very small size.

In recent years, considerable researches on CNTs sensors have been carrying on for the exploration of novel sensing materials. In 2000, Kong et al. [48] first applied the SWCNT to prepare gas sensors for detecting the mixture of NO_2 and NH_3, and they found that the conductivity of sensors after adsorbing NH_3 can reduce two orders of magnitudes and improve three orders of magnitudes in terms of NO_2. In that case, the sensors can be regarded as having a relatively

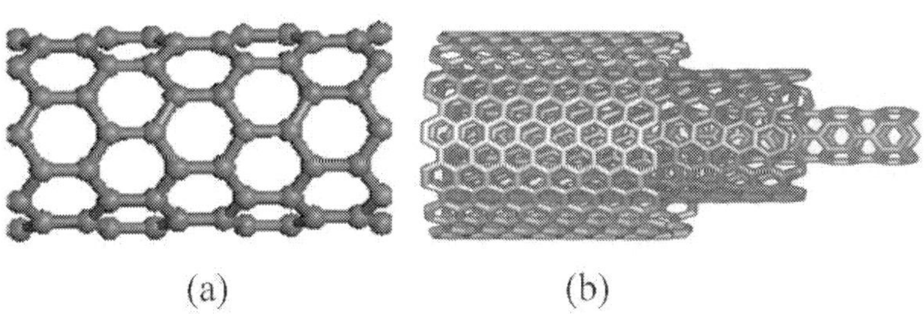

Figure 2. Structure sketch of CNTs: (a) single-wall nanotubes, (b) multi-wall nanotubes.

high selectivity in complex gas environment. Subsequently, Kong et al. [49] successfully prepared the N-modified SWCNTs with inconsecutive Pt metal thin film, which has better sensing property and quicker recovery property to H_2, demonstrating that this semi-conductivity single-wall CNTs sensors have better sensitivity, selectivity, and recovery characteristics than before. Dai et al. [6] developed a novel type of CNTs sensors, which were also made by SWCNT that have the semi-conductivity property, and they studied the change of electric property with the import of gases. The results showed that the gas responses of NO_2 and NH_3 are both good. Qi et al. [50] introduced a type of gas sensors using SWCNTs, which can detect NO_2 with the minimum content to 1 ppb.

Varghese et al. [51] proposed two kinds of means to prepare CNTs sensors. The first one is to cover a layer of CNTs-SiO_2 thin film on the flat interdigital capacitor, named capacitance sensor; the other one is to carve a crooked SiO_2 groove on a Si substrate and then to grow CNTs on the SiO_2, called resistive transducer. It has been proved that these two types of sensors are both sensitive to NH_3, presenting the liner change. Modi et al. [52] employed CNTs arrays and developed micro gas-sensing device using Thompson discharge characteristic that can sensitively detect the content of atmosphere gases.

Robinson et al. [53] of the America Marine laboratory designed a capacitance gas sensor based on CNTs, where the interdigital electrodes and the SWCNT that distributes in the interdigital electrodes are acted as a counter electrode of the capacitance, the 30nm SiO_2 layer acted as insulating layer between two poles, low-resistance silicon acted as another counter electrode. The experimental results showed that these sensors have quick response and short recovery time to NO_2, NH_3, and dimethyl methylphosphonate (DMMP). Pulichel M. hjayan and Nikhil Koratkar in Rensselaer Polytechnic Institute successfully developed the micro gas-sensing samples, which can sensitively quantify and qualitatively analyze varying gases in the atmosphere.

When implementing direct voltage on the gas sensors, the low voltage would generate strong electric field on the CNTs, and therefore form the dielectric breakdown condition. The experimental results reveal that the voltage values vary obviously with the difference of the type of gas, so that it can be qualitatively analyzed. The analyzable gases are extensive, ranging from Ar to He as well as some inert gases. Furthermore, it has been proved that the generated electricity values present direct proportion to the logarithm of the concentrations, which indicates the gas content could be quantitatively analyzed.

Zhang et al. [54] of Xi'an Jiaotong University as well as Bondavalliet al. [55] had performed an in-deep research to Thompson discharge type CNTs gas sensor. Xi Li also conducted primary study on CNTs film sensors. Zhang et al. [56] of Chongqing University also performed related studies on the electric properties of CNTs.

In terms of sensing mechanism of CNTs, many calculations based on first principle theory have been carried out. Goldoni et al. [57] deemed that the reasons SWCNTs are sensitive to O_2, CO, H_2O, and N_2 could be attributed to the combined induction of surfactant, lauryl sodium sulfate and contaminant come from NaOH, or the chemical adsorption between the defect zone of the CNTs and O_2.

Jing Li [58] considered that two sensing mechanisms are existed in the adsorption process of SWCNTs: (i) the direct charge transfer between single-wall CNTs and the acceptor or receptor, inducing the change of semi-conductivity CNTs in Femi energy, which further results in the change in conductivity, named in-tube adjust and (ii) inside the SWCNT existing the adsorption points between targeted molecules and SWCNT, leading to the charge transfer, contributing to the change of conductivity, named inter-tube adjust.

Zhou et al. [59] who applied density functional theory calculated the effect of B- and N-doped SWCNT on adsorption to H atom and H_2 molecules. Owing to the complexity of porous materials and diversity of doping substance of CNTs, the gas-sensing mechanism of CNTs still remains in qualitative or half qualitative stage, which also needs further studies both experimental and theoretical.

To improve the sensitivity and selectivity of the CNTs sensors, a large number of modified methods such as chemical doped, molecules doped, molecular coating, as well as mechanical deformation [60–69] were introduced by scholars to modify the CNT, and acquired the desirable results. The variety of chemical doped and doped materials contributes to the selectivity to various gas adsorption. For example, the B-doped and N-doped make the intrinsic CNTs become P-model and N-model semiconductor, improving the density of the carrier so that make the charge transforms much easier between gas molecules and CNTs. Through doping Au, Pt, Pd, Ir, and the other expensive metal nanoparticles, the activation energy of chemical adsorption for gases can be reduced; at the same time, these expensive metal nanoparticles become the core of the catalytic activity, so that can effectively enhance the sensitivity, selectivity, and response time of the sensors.

3. Theoretical analysis of modified CNTs-based gas sensor

3.1. Gas-sensing properties of CNTs-OH gas sensor

3.1.1. Calculation details

The simulation of quantum mechanics is realized by the Dmol^3 of Material Studio software, which is developed by an American company of Accelrys. The PW91 function of the generalized gradient approximation (GGA) was employed for the exchange correlation of electrons. P polarized function is used for modified hydroxyl (OH)-wall (8, 0) SWNT in presence of gas molecule adsorption density functional calculations. Previous theoretical calculations [70] show that the generalized gradient approximation (GGA) method can accurately describe the geometric structure and electronic structure of CNTs, and the process of interaction with molecules. To avoid the interaction between the nanotubes, we designed a large lattice of 20 Å × 20 Å × 85 Å, and use the periodic boundary conditions. In a superlattice, the SWNT-OH is made up with 64 C atoms and the -OH which modified in the sidewall of CNTs. The initial action distance between gas dissolved in oil-filled transform and SWNT-OH can be set to 0.15 nm. All atoms are calculated by the atomic potential, self-consistent field convergence value is set to 10×10^{-5}. Literature [70] shows we can get more accurate calculation results on CNTs (8, 0) brillouin zone 2 K points. All the calculation procedures completed on the Dmol^3.

Before each calculation, the first stage is to optimize the SWNT-OH of the superlattice and the isolated typical dissolved gases to get their stable configuration. And CO, H_2, CH_4 C_2H_4, and C_2H_2 are chosen as the typical dissolved gases to be detected by SWNT-OH in this part. Then let the CO, H_2, CH_4, C_2H_4, and C_2H_2 molecule in various passible ways, respectively, to approach the O and H atom of the -OH of the tube wall to make atomic optimization, and form the oil-dissolved gases molecule SWNT-OH system preliminarily. Finally, this system is unconstrained optimized to find the stable configuration and calculate its electronic properties.

3.1.2. Calculate results

In **Figure 3**, it is the SWNT-OH stable configuration after geometry optimization. In **Figure 4(a)–(e)**, they are CO, C_2H_2, H_2, CH_4, and C_2H_4 stable configuration after geometry optimization. And in **Table 3**, is the calculation results of SWNT-OH respectively absorb the CO, H_2, CH_4, C_2H_4, and C_2H_2. In **Figure 5**, are the most stable configuration of after interaction between the geometrically optimized gas molecules and SWNT-OH. The unit of structure parameter is Å. The brackets correspond to adsorption energy which unit is eV. The charge transfers Q_T between oil-dissolved gases molecule and SWNT-OH are shown in **Table 3**.

In order to determine the most stable geometry configuration of the system of oil-dissolved gases molecule and SWNT-OH, we designed the different initial configuration. In other words, let the different atoms of CO, H_2, CH_4, C_2H_4, and C_2H_2 molecule with the same ini-

Figure 3. (8, 0) SWNT-OH.

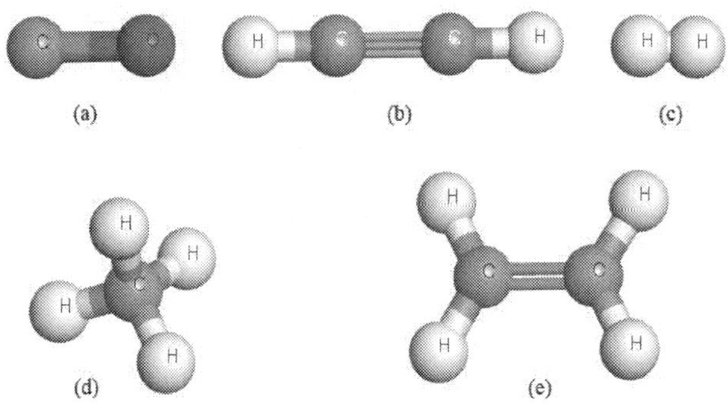

Figure 4. Optimized supercell structures for (a) CO, (b) C_2H_2, (c) H_2, (d) CH_4, and (e) C_2H_4.

tial distance (1.5) approach the O and H atom of -OH to optimize. And in order to evaluate the adsorption energy between molecules and SWNT-OH, we calculated their adsorption energy E_b. E_b is defined as $E_b = E_{(B + SWNT\text{-}OH)} - E_{(SWNT\text{-}OH)} - E_{(B)}$. In this formula, $E_{(B + SWNT\text{-}OH)}$ is the total energy of molecular adsorption on the surface of SWNT-OH; $E_{(SWNT\text{-}OH)}$ and $E_{(B)}$, respectively, are the energy of SWNT-OH and single molecule. If $E < 0$, the adsorption process is an exothermic process, the adsorption can be occurred spontaneously. In order to characterize the electrical conductivity of SWNT-OH, we calculated the charge transfers Q_T between molecule and SWNT-OH. The amount of charge transfer is able to provide the important information of the electronic response of the system. QT is defined as the charge transfer between SWNT-OH and single molecule [71].

As shown in **Table 3**, all of adsorption energy between CNT-OH and each oil-dissolved gas molecules are less than 0.6 eV. Therefore, the interaction between CNTs-OH and each oil-dissolved gas molecules is physisorption because chemisorption energy should be larger than 0.6 eV. As shown in **Figure 5** and **Table 3**, the value of charge transfer between CO and CNTs-OH (0.052 au) is nearly six times the value of that between H_2 and CNTs-OH (0.012 au) though

Structural system	Graphic	Adsorption energy (eV)	Charge-transfer (au)	Interacting distance (nm)
H_2 + SWNT-OH	**Figure 5(a)**	−0.25	0.012	0.2211
CO + SWNT-OH	**Figure 5(b)**	−0.22	0.062	0.2159
CH_4 + SWNT-OH	**Figure 5(c)**	−0.36	0.017	0.2409
C_2H_4 + SWNT-OH	**Figure 5(d)**	−0.49	0.052	0.2413
C_2H_2 + SWNT-OH	**Figure 5(e)**	−0.59	0.068	0.2409

Table 3. Calculated binding energy, net charge transfer, and interacting distance.

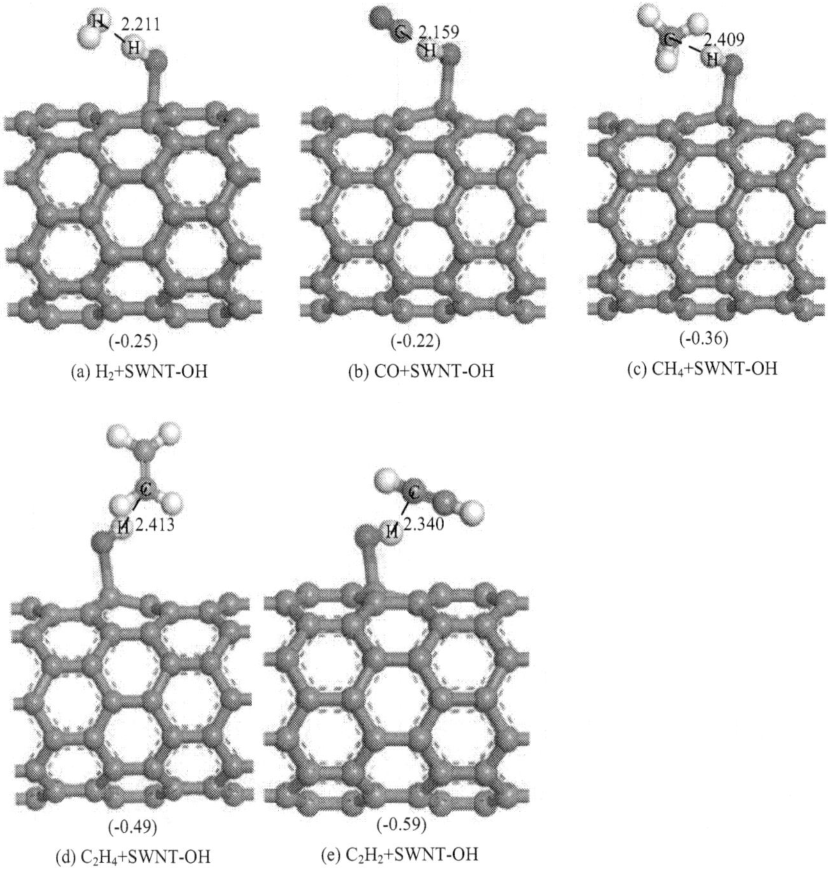

Figure 5. The most stable configurations of H_2, CO, C_2H_4, and C_2H_2 interacting with SWNT-OH after optimization, respectively.

the interaction distance and adsorption energy between H_2 and CO and CNTs-OH are almost the same. Then CNTs-OH can be used to detect CO due to the strong sensitivity if there is no organic gas in oil-dissolved gas molecules. Comparing with inorganic gas, the average adsorption energy between organic gas (CH_4, C_2H_4, C_2H_2) is about 0.48 eV, which is two times the average adsorption energy (0.23 eV) between inorganic gas (CO, H_2) and CNTs-OH. In addition, the charge transfer (0.046 au) between organic gas and SWNT-OH is far more than that between the inorganic gas value (0.037 au) between inorganic gas and SWNT-OH. This is because the hydroxyl modification on the surface of SWNT enhances its interaction to organic gas molecules due to the activation of hydroxyl. Thus, SWNT-OH is more sensitive to organic gas in oil. If we only consider the organic gas: CH_4, C_2H_4, and C_2H_2, the adsorption energy decreases in order: C_2H_2 (0.59 eV) > C_2H_4 (0.49 eV) > CH_4 (0.36 eV), and charge transfer decreases in order: C_2H_2 (0.068 au) > C_2H_4 (0.052 au) > CH_4 (0.017 au) as shown in **Table 3**. With the increase of C—C covalent, it leads to the increase of adsorption energy, resulting in the high sensitivity to C_2H_2. Hence, CNTs-OH can be used to detect C_2H_2 component in oil-dissolved transformer.

In order to evaluate the influence of oil-dissolved gas to the change of conductivity during the adsorption process, density of states (DOS) is calculated as shown in **Figure 6**. On comparing the calculation results shown in **Figure 6(a)–(f)**, gas adsorption narrows down the DOS at fermi level. Upon inorganic gas: H_2 and CO adsorption shown in **Figure 5(b)** and **(c)**, it is found that the energy gap of DOS around fermi level for CO + CNTs-OH is smother and narrower than that of H_2 + CNTs-OH adsorption system, signifying the increase of conductivity after CO adsorption. The result is also in consistence with the results that the charge transfer in CO + CNTs-OH system is larger than that of H_2 + CNTs-OH. Upon organic gas: CH_4, C_2H_4, and C_2H_2 adsorption shown in **Figure 6(d)–(f)**, the DOS of organic gas molecules adsorbed SWNT-OH system at fermi level is greater than that of inorganic adsorbed SWNT-OH system, indicating the strong interaction between organic gas molecules and SWNT-OH comparing with that of inorganic adsorption, which is also consistence with the results in **Table 3**. And the DOS for C_2H_2 + CNTs-OH system around at fermi level is obviously larger than that of other gas adsorption systems, thus SWNT-OH is most sensitive to C_2H_2 gas. Therefore, SWNT-OH can be used to detect C_2H_2 in oil-filled transformer.

3.1.3. Conclusion

In this study, density functional theory has been used to study the adsorption properties of hydroxyl-modified CNTs (CNTs-OH) upon gases dissolved in oil-filled transformer. According to the calculation results of first-principles calculations, the adsorptions to all of the gases are physisorption, which leads to the change of geometric and electronic structures. The adsorption energy to organic gases is bigger than inorganic gases, especially reflecting in the great adsorption energy to C_2H_2. Therefore, we conclude that SWN-OH can be chosen as gas sensor to detect C_2H_2 gas dissolved in oil-filled transformer.

3.2. Gas-sensing properties of Ni-CNTs gas sensor

3.2.1. Calculations details

The results showed that different oil-dissolved gases have different responses on the Ni-CNT sensor. To further understand the sensing mechanism, we established a properly simplified model to calculate and analyze the adsorption properties of the supports (CNTs and Ni-CNTs) to the gases. Ni-substituted CNTs and typical oil-dissolved gases were constructed to simulate the sensor in this part. C_2H_2, C_2H_4, and C_2H_6 are chosen as the target measured gases due to the specific sensitivity and selectivity of Ni-CNT sensor.

Totally optimized geometries and related properties of the configurations were carried out by Design for Testability (DFT) calculations in the generalized gradient approximation using the

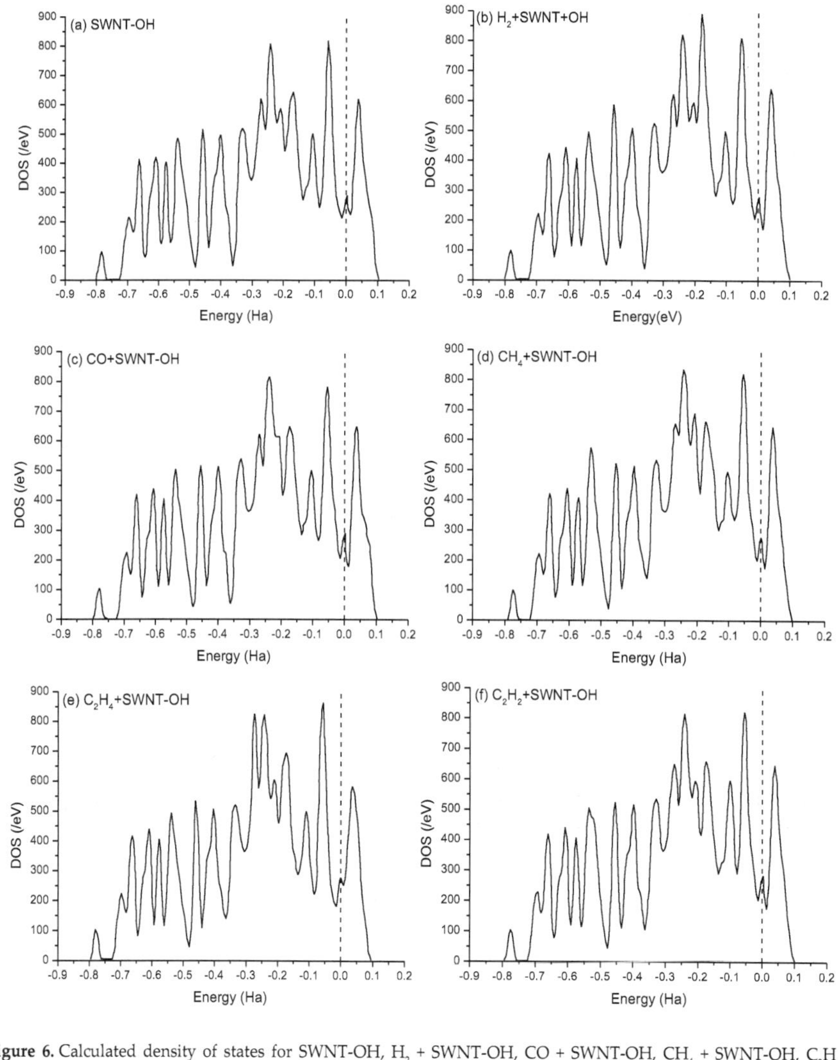

Figure 6. Calculated density of states for SWNT-OH, H_2 + SWNT-OH, CO + SWNT-OH, CH_4 + SWNT-OH, C_2H_4 + SWNT-OH, C_2H_2 + SWNT-OH.

Dmol[3] model with double-numerical polarized basis sets. The whole calculations were performed using the Perdew-Burke-Ernzerhof (PBE) DFT. The geometrical structures are shown in **Figures 7** and **8**.

3.2.2. Results and discussion

The spontaneity of these interactions can be described by the adsorption energy E_{ads}, which is defined as follows:

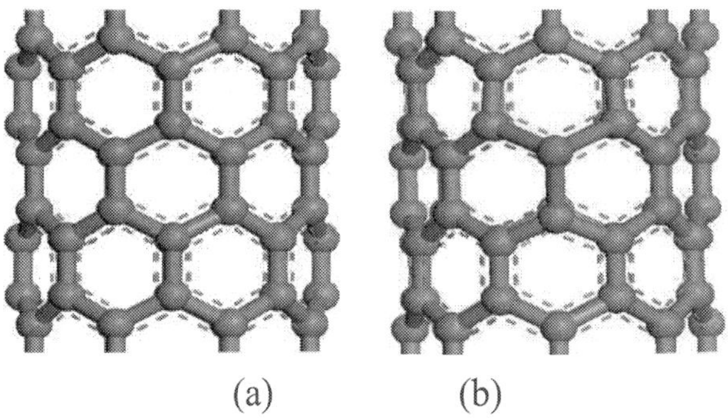

Figure 7. Geometrical structures of CNTs and Ni-CNTs. (a) CNTs, (b) Ni-CNTs.

$$E_{ads} = E_{(gas/support)} - E_{(support)} - E_{(gas)} \tag{1}$$

where $E_{(gas/support)}$ is total energy of adsorbed system after adsorption of gas molecule, $E_{(support)}$ is the energy of support without gas molecules adsorption, $E_{(gas)}$ is the energy of individual gas molecule. $E_{ads} < 0$ represents that adsorption process is exothermic and spontaneous.

Mulliken population of gas molecules and support were calculated, respectively, so that the charge distribution of the system can be obtained in the adsorption process. Charge transfers Q_T was defined as charge variation before and after adsorption of gas molecules. The values of adsorption energies were negative showed that the adsorption was exothermic. And in this CNT system, the values of the energy were small (<0.6eV), and as follow order: $C_2H_2 > C_2H_4 > C_2H_6$. The charge transfers close to zero show that the adsorptions of gas molecules on the surface of CNTs are physisorption. And the sensitivity of adsorption is as follows: $C_2H_2 > C_2H_4 > C_2H_6$.

Table 4 shows the adsorption energy and charge transfers of the Ni-CNT system and the CNT system. Compared with the CNT system, the doped Ni effectively improved the electronic structure and sensitivities of CNTs. The adsorption energies of C_2H_6 are the lowest, and C_2H_2 and C_2H_4 are 8.7 and 4.6 times larger than that of C_2H_6, respectively. The values of charge transfers of three gases are as follows: $C_2H_2 > C_2H_4 > C_2H_6$. Both the adsorption energies and charge transfers of C_2H_2 are highest. Thus, Ni-CNTs have the highest sensitivity to C_2H_2, which is similar to that of the CNTs.

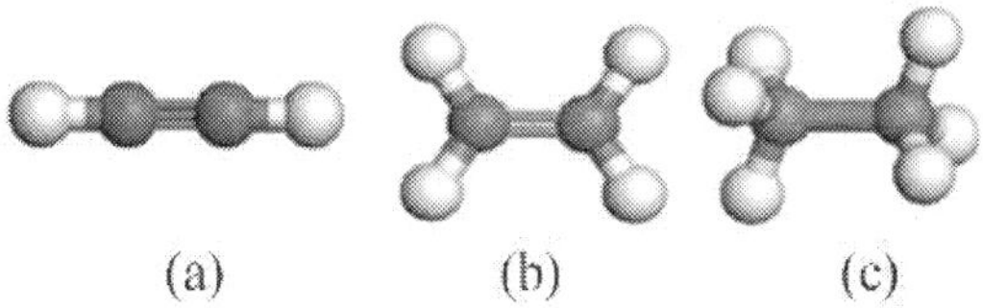

Figure 8. Geometrical structures of oil-dissolved gases (a) C_2H_2, (b) C_2H_4, (c) C_2H_6.

	E_{ads} (eV)	Q_T (e)
C_2H_2-CNTs	−0.3265	0.006
C_2H_4-CNTs	−0.2814	0.003
C_2H_6-CNTs	−0.0458	0.002
C_2H_2-Ni-CNTs	−1.7412	0.091
C_2H_4-Ni-CNTs	−0.9246	0.069
C_2H_6-Ni-CNTs	−0.1994	0.043

Table 4. Adsorption energy and charge transfer.

In summary, the sensitivity of the CNT sensor for the gases is as follows: $C_2H_2 > C_2H_4 > C_2H_6$, and the doped Ni can improve the sensor sensitivity.

The molecular orbit theory was calculated to obtain the highest occupied molecular orbital (HOMO) energy and the lowest unoccupied molecular orbital (LUMO) energy of the three gas molecules and the supports. The analysis of HOMO and LUMO and related energy gap are able to determine whether charges can easily transform between gases and adsorbent or not. $E_{L\text{-}H}$ is calculated based on the following equation:

$$E_{L\text{-}H} = E_{LUMO} - E_{HOMO} \tag{2}$$

A small $E_{L\text{-}H}$ corresponds to an easy transferred charges between the orbitals and to a good conductivity for this material. The calculated E_{HOMO}, E_{LUMO}, and $E_{L\text{-}H}$ are depicted in **Table 5**. The $E_{L\text{-}H}$ of CNTs and Ni-CNTs are 0.6911 and 0.5470 eV, respectively. The Ni dopant reduces the $E_{L\text{-}H}$ of 0.1441 eV so that enhances the conductivity of the carbon nanotube. After adsorption, the frontier orbital energies of the adsorption configurations are increased and $E_{L\text{-}H}$ changed as well, their values are in order as follows: C_2H_2-Ni-CNTs < C_2H_4-Ni-CNTs < C_2H_6-Ni-CNTs. Therefore, conductivities of adsorbent in the adsorption systems are in order as follows: C_2H_2-Ni-CNTs > C_2H_4-Ni-CNTs > C_2H_6-Ni-CNTs.

In the process of Ni-CNTs adsorption, the conductivity change of the adsorption system shows that C_2H_2 is the highest, while C_2H_6 is the lowest. The change of resistance shows

Adsorption type	E_{HOMO} (eV)	E_{LUMO} (eV)	$E_{L\text{-}H}$ (eV)
CNTs	-4.5606	-3.8695	0.6911
Ni-CNTs	-4.9797	-4.4327	0.5470
C_2H_2-Ni-CNTs	-4.5906	-4.1606	0.4300
C_2H_4-Ni-CNTs	-4.6940	-4.2477	0.4463
C_2H_6-Ni-CNTs	-4.7593	-4.1933	0.5660

Table 5. Molecular frontier orbital energy and orbital energy differences.

the same changing characteristics. This result indicates that the sensitivity of Ni-CNTs is as follows: $C_2H_2 > C_2H_4 > C_2H_6$. This finding is consistent with the results based on gas-sensing experiments.

3.2.3. *Analysis of theoretical calculate*

In this chapter, to detect oil-dissolved gases in a transformer, the research's work includes theoretical and experimental studies on a Ni-CNT sensor. This study focuses on the response and mechanism of gas sensing. The results of gas-sensing experiment are consistent with the simulation.

Charge redistribution between the surface and the adsorbed molecules results in changes in the electronic structure and conductivity. Higher charge transfer results in greater conductivity changes. The values of transfer charges calculated based on DFT are shown in **Table 4**. Compare with other two gases, the transfer charges of C_2H_2 are the highest, and C_2H_2 has the highest response on the gas sensor. In addition, the orbital theory results are consistent with the charge transfer analysis and experimental results. Thus, in this chapter, the theoretical analysis results are consistent with the experimental results, and the sensitivities of the three gases on Ni-CNTs are as follows: $C_2H_2 > C_2H_4 > C_2H_6$. Moreover, as the C_2H_2 concentration increases, the response time becomes shorter. High gases concentration leads to fast sensor response.

Previous researches signified that gas-sensing properties can be enhanced by metal doping. The transition metal is rich in d-electrons and has empty orbits, and the small gas molecules can be strongly combined with the metal when adsorbed on the surface. In this chapter, nickel ions are the transition metal divalent cations used, which make nickel ions more accessible to the internal tubes in the capillary. Moreover, due to the coordination unsaturation of the surface atoms of nickel ions, the surface active sites of CNT increase and the catalytic activity is greatly enhanced. In general, the order of the chemical adsorption capacity of the transition metal to the gas is as follows: $O_2 > C_2H_2 > C_2H_4 > CO > H_2 > CO_2 > N_2$. The results of this paper are consistent with this order, which indicating that the doped Ni increases the chemical adsorption of the gas molecules.

3.2.4. *Conclusion*

a. The adsorption between three gases and intrinsic CNTs is physical adsorption, and the order of adsorption sensitivity is as follows: $C_2H_2 > C_2H_4 > C_2H_6$.

b. The adsorption sensitivity of the Ni-CNTs was consistent with that of the CNTs, doping Ni improves the conductivity of the CNT, and the adsorption of the three gases becomes easier on Ni-CNTs.

c. Due to the coordination unsaturation of the surface atoms of nickel ions, the surface activity sites of CNTs increase and the catalytic activity is greatly improved. Doped Ni improves the ability of the tube to adsorb gas molecules.

d. When a low concentration (1–10 μL/L) of C_2H_2 is detected, the relative change in sensor resistance *R*% and gas concentration satisfies a certain linear relationship, indicating that the developed sensor can detect low gas concentrations.

4. Experimental analysis of modified CNTs-based gas sensor

4.1. Preparation of CNTs sensors

4.1.1. Arc discharge method

It is the first and most used method to prepare CNTs. The main processes are: (a) keep a certain pressure of inert gas or hydrogen in vacuum vessel and (b) choose graphite (with catalyst: nickel, cobalt and iron, etc.) as electrode. The graphite is consumed by evaporation at anode during the arc discharge process, and CNTs are received by depositing at cathode. Ebbsen and Ajayan [72] successfully prepared gram order weight of CNTs under nitrogen gas condition, and then this method is widely adopted. In 1994, Bethune introduced catalyst for arc reaction, reducing the reaction temperature and enhancing the productivity of CNTs. In 1997, Journet et al. [73] used catalysts for synthesizing single CNTs under helium condition. Mingliang et al. [74] studied the influence factors to CNTs prepared by DC arc discharge method: (a) inert gas pressure will affect the diameter and length of CNTs. (b) How much the adhesion of particles? (c) Oxygen and water vapor will lead to defects in CNTs, and it is unable to separate and purify after sintering together. (d) Current and voltage will affect the yield and production rate of CNTs, but length to diameter ratio of graphite does not affect the generation of CNTs.

4.1.2. Catalytic cracking method

Catalytic cracking method, also known as chemical vapor deposition, prepares CNTs through cracking hydrocarbons or carbon oxides with the help of catalyst. The basic preparation processes are: (a) mix the organic gases (such as acetylene and ethylene) with certain proportion of nitrogen gas in quartz tube. (b) CNTs grow on the surface of catalyst under certain temperature when the carbon source flow past and pyrolysis onto the surface of catalyst, and pushing forward the small catalyst particles [75]. (c) The growth of CNTs ends till all of the catalyst particles were coated with graphite layer. The advantages of the method are: easy to control the reaction process, simple equipment, low raw material cost, easy to produce the product in large scale, and the high productivity. The disadvantages are: too much CNTs layers, poor graphitization, exist crystalline defects. Theses disadvantages have great adverse influence on the physical and chemical properties of CNTs.

4.1.3. Laser evaporation method

Laser evaporation method prepares CNTs by illuminating the graphite target that contain metal catalyst. Then the vapor mix with carbon source and deposit on the surface of substrate and the wall of reaction chamber. Smalley et al. received SWCNTs after adding a certain amount of catalyst to the electrode during preparing C_{60}. After improving the method, Thess et al. [76] successfully fabricated amount of SWCNTs. Under the condition of 1473 K, the graphite target with Ni/Co catalyst particles was irradiated by double pulse laser with 50 ns, receiving the high quality SWCNTs bundles.

4.1.4. Low temperature solid state pyrolysis method

Low temperature solid state pyrolysis prepares CNTs through intermediate. First, the nanometer level silicon nitride (Si_2C_2N) ceramic intermediate was prepared. The nanoceramic intermediate is then placed in a boron nitride crucible, which is heated in a graphite resistance furnace to decompose it with nitrogen gas as the protective gas. After 1 h, the nanointermediate powder begins to paralyze, and the carbon atoms migrate to the surface. A high proportion of CNTs is obtained with amount of silicon nitride powder in the surface pyrolysis products. The advantage of the low temperature solid state pyrolysis method is the repeatable production, which is beneficial for large-scale CNTs production.

4.1.5. Polymer pyrolysis method

The method prepares CNTs by decomposition of hydrocarbons precursor (such as acetylene and benzene) at high temperature. Cho et al. [70] prepared CNTs by heating the polymer obtained from citric acid and glycol after polyesterification under 400°C for 8 hours. The CNTs were synthesized by using metal Ni as catalyst in the temperature ranged from 420 to 450°C and under H_2 atmosphere. Under the 900°C and Ar-H_2 atmosphere conditions, Sen et al. [77] obtained CNTs by pyrolyzing ferrocene, nickelocene, and cobaltocene. These metal compounds not only provide carbon source after pyrolysis but also provide the catalyst particles. The growth mechanism of the method is similar to the catalytic cracking method.

4.1.6. Ion (electron beam) radiation method

In a vacuum furnace, carbon is evaporated by ion or electron discharge and deposit on the condenser. Chernozatonskii et al. [78] synthesized CNTs with diameter range from 10 to 20 nm and high alignment by evaporating the graphite coated on the surface of substrates. Yamamoto et al. [79] got CNTs with diameter range from 10 to 15 nm by irradiating amorphous carbon with argon ion beam under high-vacuum environment [80].

4.1.7. Flame synthesis method

Flame synthesis method utilizes the heat, produced by burning methane and a small amount of oxygen, and imports hydrocarbons and catalysts at temperature of 600–1300°C to synthesize CNTs. The CNTs prepared by this method have the disadvantages of low crystallinity and large amount of amorphous carbon. There is still no definite explanation for the growth mechanism of CNTs nanostructure by flame method. Richter et al. [81] found SWCNTs that attached with a large amount of amorphous carbon from carbon black after burning the mixture of acetylene, oxygen, and argon gases. Das Chowdhury et al. [82] found nanometer tubular CNTs by detecting carbon black after burning the mixture of benzene, acetylene, ethylene, and oxygen gases.

4.1.8. Solar energy synthesis method

The CNTs are received from the condensation of high temperature (3000 K) mixture vapor of graphite and metal catalyst that heated by focusing the sunlight. This method is initially used for buckyballs production, then adopted for CNTs synthesis since 1996. Laplaze et al. [83] synthesized that the CNTs and SWCNTs use this method.

4.1.9. Electrolysis method

The preparation of CNTs by electrochemical method is a novel technique. This method adopted graphite electrode (electrolytic cell as anode) and obtained carbon nanomaterials by electrolyzing molten alkali halide salts (such as LiCl) under a certain voltage and current with the protection of air or argon gases at about 600°C. The products include packaged or not packaged CNTs and carbon nanoparticles, and the form of carbon nanomaterials can be controlled by changing the process conditions of electrolysis. Goldoni et al. [57] found that CNTs can directly grow on the surface of n type of (1 0 0) silicon electrode in solution of acetylene/ammonia. Hsu et al. [84] synthesized nanotubular and onion-like CNTs under argon environment by using molten alkali metal halide as electrolyte and graphite as electrode. Hui et al. [85] successfully synthesized CNTs and carbon nanowires using LiCl and LiCl + $SnCl_2$ as molten salt electrolyte.

4.1.10. Other methods

Stevens et al. [86] got CNTs by using an exothermic reaction between cesium and nanoporous amorphous carbon in the low temperature of 50°C. Chernozatonskii et al. [87] found the fullerene and CNTs at the micro holes of Fe_2Ni_2C, Ni_2Fe_2C, and $Fe_2Ni_2Co_2C$ alloy prepared by powder metallurgy method [88]. Kyotani et al. [89] first pyrolyze and deposit carbon on the wall of anodic alumina model (with nanometer trench) under 800°C. Then hollow CNTs with open-end on both sides after removing the anodic alumina membrane by hydrofluoric acid. Matveev et al. [90] synthesized CNTs using liquid nitrogen solution of acetylene at 233 K by electrochemical method. It is the lowest temperature ever reported to synthesize CNTs.

4.2. Gas-sensing properties of mixed acid-modified CNTs gas sensor

4.2.1. Sample preparation

This chapter used CNTs which were made by chemical vapor deposition method. Tube diameter is 20–30 nm, with a length of 10–30 μm, purity > 95%, the catalyst residue (ash) < 1.5 wt%, and multi-walled structure. Around 0.1 g of CNTs was placed in an appropriate amount of anhydrous ethanol, then adding surface active agent. Afterward, it had been scattered by ultrasonic oscillator with 2 hours in order to obtain a moderate concentration of CNTs solution, which was set as the sample I. Similarly, another 0.1 g of CNTs was immersed in 50 mL mixed acid solution, which was initially prepared by mixing concentrated sulfuric acid and concentrated nitric acid with the volume ratio of 3:1. The solution was put in ultrasonic oscillator for about 2 hours of dispersion. And then the solution was diluted with deionized water, and filtrated by filter membrane with an aperture of 0.22 μm. This progress should be conducted repeatedly until the diluted solution becomes neutral. The finally collected sample is named as sample II.

4.2.2. Sensor preparation

The printed circuit board was used to make the substrate of CNTs sensors. The surface of substrate was etched by copper to generate interdigital electrodes. Copper foil has the thick of 30 μm, with 0.5 mm electrode interval and 0.5 mm line width, as shown in **Figure 9**.

Once moderate concentration solution of the sample I was obtained, the micro-scale parts were spread on the space between the interdigital electrodes, and then placed in a drying oven at 80°C to dry it. Uniform dense of MWNTs film with smooth surface was able to be prepared with repeated operations. The obtained sample is named as the sensor I, as shown in **Figure 9(b)**. Take appropriate sample II placed in anhydrous ethanol, and then after ultrasonic dispersion for 10 min to get moderate concentration of the suspension. Sensor II was obtained in the same way. These two-dimensional CNTs films generated from the deposition of the one-dimensional CNTs have so many structural defects, making it possess specificity of electrical properties.

4.2.3. Detection of dissolved gases in oil test using MWNTs sensor

Device for detection of dissolved gas in transformer oil by CNTs-based sensors is shown in **Figure 10**. Prepared CNTs sensors were initially placed in the test device. It was a sealed chamber

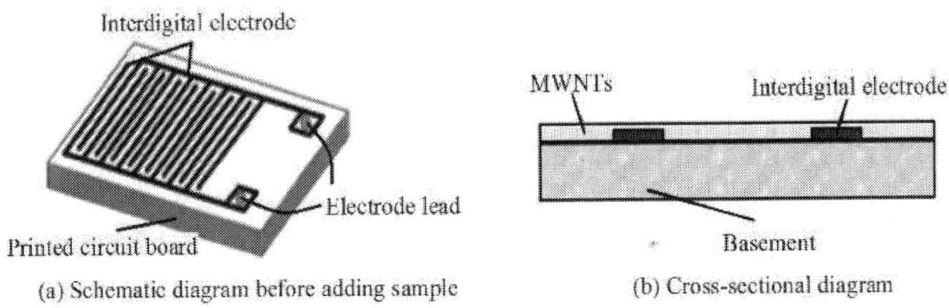

Figure 9. The geometric sketch of CNTs sensor.

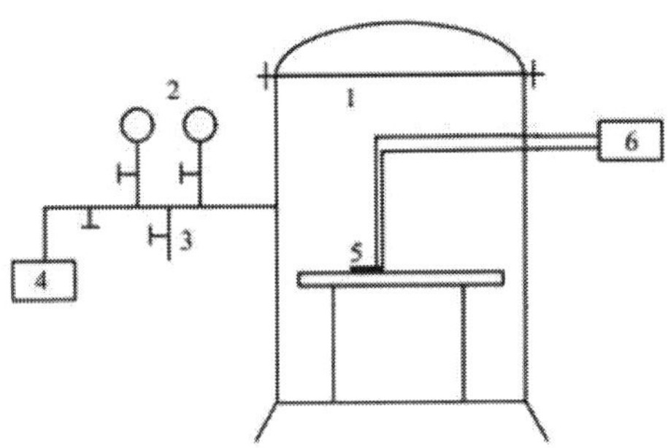

Figure 10. Detection test device for the CNTs sensor adsorbing gases dissolved in transformer oil. 1: sealed metal can, 2: vacuum gauge and pressure gauge, 3: intake valve, 4: vacuum pump, 5: CNTs sensors, 6: impedance analyzer.

designed to perform this experiment. Then the sensor was connected with the impedance analyzer by wire and the chamber was then sealed with a round head passing through the spherical ring with a screw and a nut.

The standard gas of CH_4 with concentration of 200 μL/L is injected through the intake valve into the test device, as shown in **Figure 10**, and use sensors I and II to detect gas response, respectively. The acquired gas response curves are shown in **Figure 11**. Based on **Figure 11**, $S\%$ indicates the relative change of resistance, R is the sensor resistance value after the interaction with injected gas, and R_0 is the resistance of the sensors in a vacuum environment.

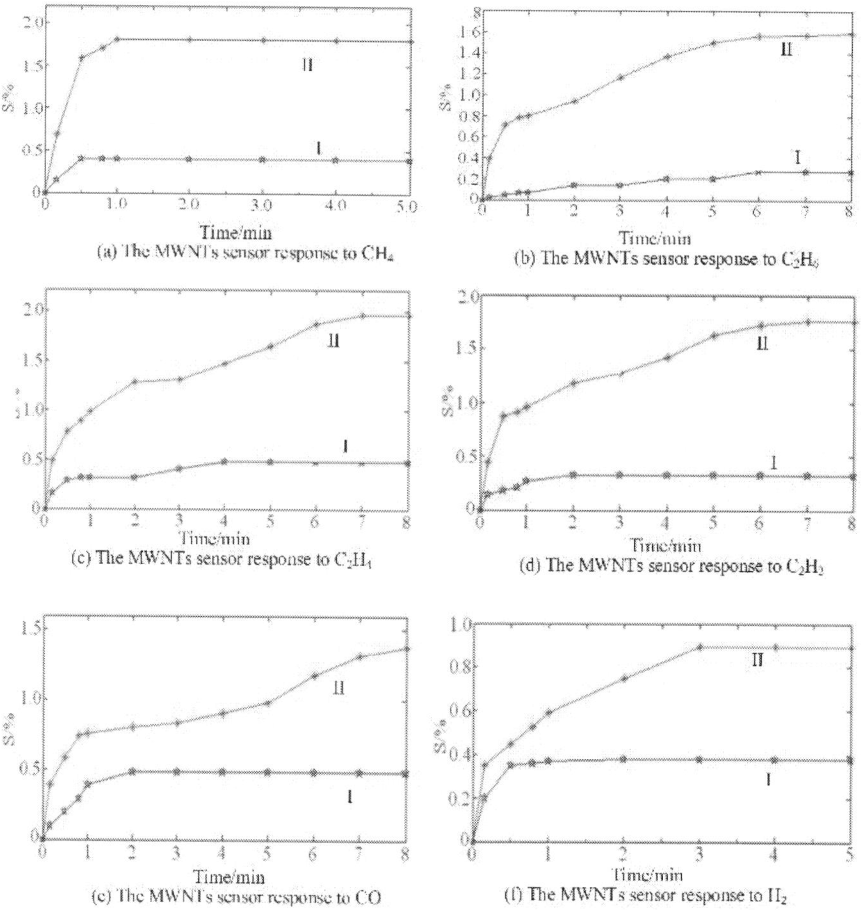

Figure 11. The MWNTs sensor response to CH_4, C_2H_6, C_2H_4, C_2H_2, CO, H_2.

In **Figure 11(a)**, curves I and II show response curve of sensors I and II, respectively. It can be observed that the changes of MWNTs thin-film sensor in resistance value are very small without chemical modification, about 0.4%; while the one conducted with chemical modification has great change with respect to resistance value, reaching to 1.8%.

After detection, a vacuum pump is used to make the device vacuumed again, according to the same method. Inject C_2H_6, C_2H_4, C_2H_2, CO, and H_2 gas into the tank, respectively, with concentration of 200 μL/L. Then detect the response curve using sensors I and II, respectively. Obtained response curves are shown in **Figure 11(b)–(f)**.

Combining **Figure 11(a)–(f)** obtains **Table 6**, the resistance value change of CNTs which had adsorb the measured gas and being chemically modified is much greater than that of the non-modified CNTs. After calculation, adsorption capacity of modified CNTs to CH_4, C_2H_6, C_2H_4, C_2H_2, CO, and H_2 increased by about 4.6, 5.9, 4.2, 5.3, 2.9, and 2.4 times, respectively. It can be seen that chemical modification contributes great affection upon the electrical properties of MWNTs.

The CNTs film is regarded as a connection of many disordered CNTs or CNTs, among which there are considerable series-parallel paths. The high-resistance samples contain more series path than parallel paths, while low-resistance samples on the opposite. Gas-adsorption property is closely related to charge transfer capacity, adsorption sites, and the characteristics of gas molecules as well. Adsorption sites of gas molecules in carbon nanotubes include: the tube gaps of a bundle of CNTs, the grooves on the surface of the bundle between tubes, the inner cavity of CNTs as well as the tube surface. Recently, many researchers agree that there are two kinds of carriers in CNTs, electrons and holes, Cantalini et al. [91] and some other scholars argue that the MWNTs is a P-type semiconductor properties, namely the electron would be accepted by gas molecules after adsorption of oxidation substances, so that concentration of holes would be increased, and resistance value would be decreased; on the other hand, once reducing gases are adsorbed on the CNTs surface, their electrical resistance increases. Given that in oil, methane and other gases have certain reducibility, it can result in an increase in electrical resistance after their adsorption on CNTs surface. As analyzed, if the carbon nanotubes are pretreated with concentrated nitric acid and concentrated sulfuric acid, its length would be shortened, ports be opened, and a lot of depressions on the surface

Gas (200 μL/L)	The relative change in resistance value (%)	
	I	II
CH_4	0.39	1.81
C_2H_6	0.27	1.60
C_2H_4	0.47	1.96
C_2H_2	0.33	1.76
CO	0.48	1.38
H_2	0.38	0.90

Table 6. Relative changes of CNTs resistance value to different gases in oil.

as well as at ports be generated. Then a large number of stable functional groups such as carboxyl, hydroxyl, and carbonyl are bonded to the adsorbing sites of depression, which would increase the number of active sites for gas adsorption, contributing to better gas sensitivity.

4.2.4. Conclusion

CNTs-based gas sensors that have advantages of high sensitivity, fast response, and small size can work at room temperature. This chapter took good use of these electrical properties, introduced a multi-walled CNTs gas sensor. Laboratory mixed acid modification was employed to improve the gas-sensing properties of CNTs to dissolve gases in transformer. Results show that without modification, MWNTs sensors are insensitive to dissolved gas in transformer oil; while the modified MWNTs sensor that has many faults and contains active functional groups guarantee the good sensitivity and fast response characteristics to the dissolved gas in oil. Synthesis of CNTs sensors industrially and large-scaly to realize this purpose is hard, but it provides a novel way for this detection. In the further work, researches should focus on gas-sensing response mechanism, sensitivity, and selectivity of so-prepared CNTs. It is hopeful and promising to prepare CNTs-based sensors that have better performance for detection of dissolved gas in oil.

4.3. Gas-sensing properties of Ni-CNTs gas sensor

Due to the low growth temperature, and the atmospheric pressure during the reaction, etc., the Chemical Vapor Deposition (CVD) is widely used in the synthesis of CNTs. The sensitivity of the sensor to typical oil-dissolved gases was studied. C_2H_2, C_2H_4, and C_2H_6 are chosen as the target measured gases in consistence with theoretical calculation above.

4.3.1. Preparation of the Ni-CNT sensor

In this chapter, the purity of the CNTs is more than 95%, which diameter and length are ranged from 20 to 30 nm and 10 to 30 μm, respectively. At first, a mixed solution of concentrated sulfuric acid (98%) and nitric acid (78%) at a concentration ratio of 3:1 was arranged, then put into 0.1 g CNTs, and dispersed in an ultrasonic shaker for 60 min. Second, washed several times with deionized water until the solution became neutral and then dried at 70°C. After these two steps, the dark powder of the mixed acid-modified CNTs can be obtained.

To prepare 1 mg/mL solutions of CNTs, take appropriate amount of CNTs dissolved in anhydrous ethanol. Take 20 mg $NiCl_26H_2O$ dissolved in 50 mL 1 mg/mL solutions of CNTs. In order to obtain a uniform dispersed Ni-CNTs solution, put the beaker in an ultrasonic bath for 90 min. Using coating drops prepared the Ni-CNTs thin films on the surface of interdigital electrodes and dried at 80°C. To ensure a compact and smooth distribution of the sensing film, repeated this process.

4.3.2. Sensor response experiment

The device for detecting the gas-sensing properties is shown in **Figure 12**. The main part of the device is a steel chamber that is sealed by screws. Before the test, the pressure tightness of the device should be examined and guaranteed.

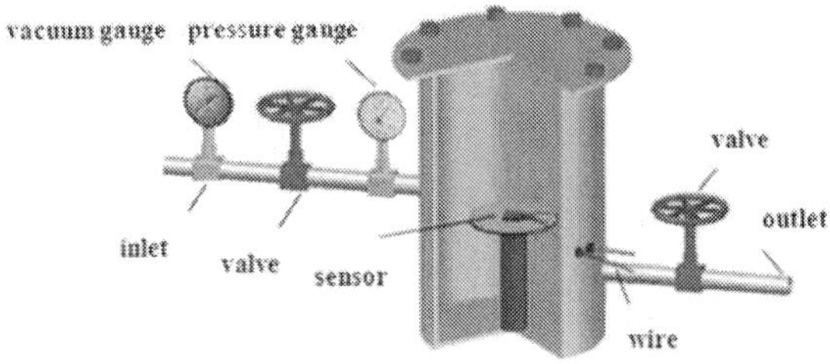

Figure 12. Geometry of the experimental equipment.

First, the sensor was put inside the chamber and connected with an impedance analyzer through wires to record the measured resistance. Second, nitrogen was passed through the chamber until the resistance of the sensor becomes stabilized. Then, different concentrations of the target gas species were injected into the sealed chamber through the inlet valve. The relative variation of the resistance was calculated as expressed:

$$R\% = (R - R_0)/R \times 100\% \tag{3}$$

where R is the sensor resistance in relevant gas and R_0 is the sensor resistance in the environment full of nitrogen. After each test, the chamber should be evacuated for the next test. All the operations in this work were performed at room temperature.

4.3.3. *Experiment result and discussion*

The gas responses of the Ni-CNTs prepared gas sensor upon the concentrations of 10 μL/L C_2H_2, C_2H_4, and C_2H_6 were detected using the method described above. The gas response curves are shown in **Figure 13**, where the horizontal axis represents time, and the vertical axis represents resistance. In order to avoid accidental factors that affect the detection results, data presented here are the results of statistical analysis preformed on 10 sensor samples instead

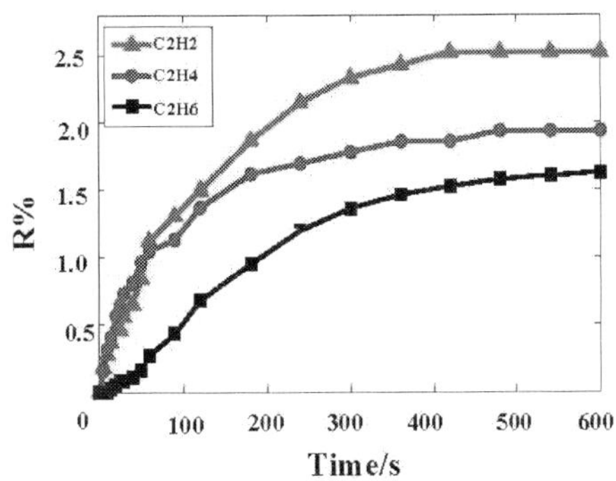

Figure 13. Ni-CNT sensor response to 10 μL/L C_2H_2, C_2H_4, and C_2H_6.

of one set. The gas sensitivity in this work is an average value. The calculated standard deviations of C_2H_2, C_2H_4, and C_2H_6 are 0.0374, 0.0288, 0.0275, respectively (data not shown).

Figure 13 shows that there is a sharp rise in the resistance of the Ni-CNT-based sensor at first when exposed to atmosphere filled with C_2H_2, C_2H_4, and C_2H_6, and then becomes stable after 400 s. It can be observed that the relative variations of the resistance for C_2H_2, C_2H_4, and C_2H_6 are nearly unchanged at 2.52, 1.95, and 1.61%, respectively. These results indicate that the Ni-CNTs sensor presents the most sensitivity to C_2H_2 under the same concentration compared with the other two gases.

4.3.4. Gas response of Ni-CNT upon different C_2H_2 concentrations

A standard value of the dissolved gas in the transformer oil is 5 μL/L. In order to meet the engineering requirements, the gas-sensitive response of C_2H_2 at concentrations of 1, 3, 5, and 10 μL/L were all tested, with related result shown in **Figure 14(a)**. The change of prepared sensors in resistance to 1, 3, 5, and 10 μL/L C_2H_2 are obtained as 0.52, 1.05, 1.18, and 2.52%, respectively. With the increasing concentration, the relevant change in resistance increases as well, and the response time is accordingly shortened. **Figure 14(b)** depicts the linear fit curve of the response and gas concentrations with the linear correlation coefficient (R^2) of 0.98. These results imply that when the C_2H_2 concentration is between 1 and 10 μL/L, the change of Ni-modified CNT in resistance meets a certain linear dependence with the gas concentration, which indicates that this material can be applied to estimate the concentration of C_2H_2 gas.

4.3.5. Reproducibility of Ni-modified CNTs

The sensor reliability is strongly depended on the reproducibility that is exhibited by the sensor material. The reproducibility of the Ni-doped CNTs sensor was evaluated by repeating the response experiments for three times. Tests were conducted according to the experimental steps described in Section 2.2. Pure N_2 was employed to accelerate desorption of gas molecules. The dynamic response transients for the Ni-doped CNTs sensors toward 10 μL/L C_2H_2 gas is depicted in **Figure 15** in order to illustrate desorption and repeatability processes. Based

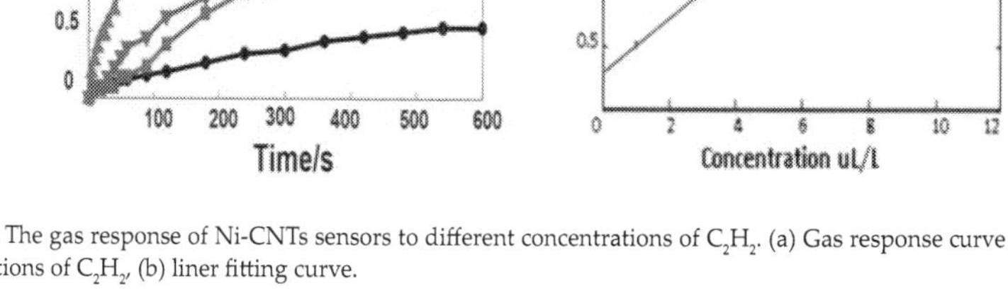

Figure 14. The gas response of Ni-CNTs sensors to different concentrations of C_2H_2. (a) Gas response curve to different concentrations of C_2H_2, (b) liner fitting curve.

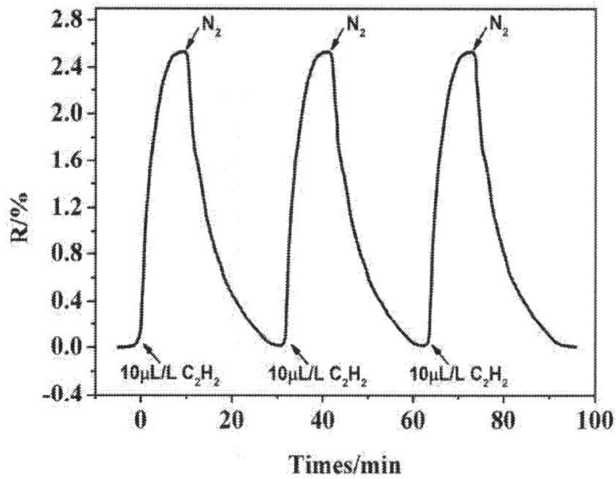

Figure 15. Reproducibility of Ni-doped CNTs sensor to 10 μL/L C_2H_2.

on this figure, one can find that the response of the material is almost constant, confirming the reproducibility of sensor material, which suggests that the Ni-CNTs-prepared sensor can be applied as a reusable sensing material for detecting oil-dissolved gases.

Author details

Ju Tang[1], Xiaoxing Zhang[1*], Song Xiao[1] and Yingang Gui[2]

*Address all correspondence to: zhxx@cqu.edu.cn

1 School of Electrical Engineering, Wuhan University, Wuhan, China

2 College of Engineering and Technology, Southwest University, Chongqing, China

References

[1] Zhang X, Xiao S, Shu N, Tang J. GIS partial discharge pattern recognition based on the chaos theory. IEEE Transactions on Dielectrics & Electrical Insulation. 2014;**21**:783-790

[2] Fofana I, Bouaicha A, Hadjadj Y, N'Cho J, Aka-Ngnui T, Beroual A. Early stage detection of insulating oil decaying. In: 2010 Annual Report Conference on Electrical Insulation and Dielectric Phenomena; October 17-20, 2010; IEEE; 2010. pp. 1-4

[3] Lu J, Zhang X, Wu X, Dai Z, Zhang J. A Ni-doped carbon nanotube sensor for detecting oil-dissolved gases in transformers. Proceedings of the Csee. 2015;**15**:13522-13532

[4] Czapowski G, Tomassi-Morawiec H. Research on the on-line measuring method for the hot spot temperatures of transformer windings. Northeastern Electric Power Technology. 2002;**18**:3491-3501

[5] Sachdev MS, Sidhu TS, Wood HC. A digital relaying algorithm for detecting transformer winding faults. IEEE Transactions on Power Delivery. 1989;**4**:1638-1648

[6] Zhang X, Gui Y, Dai Z. A simulation of Pd-doped SWCNTs used to detect SF_6 decomposition components under partial discharge. Applied Surface Science. 2014;**315**:196-202

[7] Zhang X, Gui Y, Xiao H, Zhang Y. Analysis of adsorption properties of typical partial discharge gases on Ni-SWCNTs using density functional theory. Applied Surface Science. 2016;**379**:47-54

[8] Chang T, Zhang X, Liu W, Sun C. Detecting oil dissolved gases using carbon nanotubes sensor. International Conference on High Voltage Engineering and Application. 2010;645-648

[9] Zhang X, Tie J, Zhang J. A Pt-doped TiO_2 nanotube arrays sensor for detecting SF_6 decomposition products. Sensors (Basel, Switzerland). 2013;**13:**14764

[10] Wang M. Statistic analysis of transformer's faults and defects at voltage 110 kV and above. Distribution and Utilization. 2007

[11] Zhang X, Yu L, Gui Y, Hu W. First-principles study of SF_6 decomposed gas adsorbed on Au-decorated graphene. Applied Surface Science. 2016;**367**:259-269

[12] Zhang X, Dong X, Gui Y. Theoretical and experimental study on competitive adsorption of SF_6 decomposed components on Au-modified anatase (101) surface. Applied Surface Science. 2016;**387**:437-445

[13] Zhang X, Gui Y, Dong X. Preparation and application of TiO_2 nanotube array gas sensor for SF_6-Insulated equipment detection: A review. Nanoscale Research Letters. 2016;**11**: 1-13

[14] Yue ZH, Zhong JL, Jiang JW. Application of on-line monitoring system for dissolved gases in transformer oil. Guangdong Electric Power. 2000

[15] He H, Xu X. Study on transformer oil dissolved gas online monitoring and fault diagnosis method. International Conference on Condition Monitoring and Diagnosis. 2012; 593-596

[16] Zhang X, Chen Q, Tang J, Hu W, Zhang J. Adsorption of SF_6 decomposed gas on anatase (101) and (001) surfaces with oxygen defect: A density functional theory study. Scientific Reports. 2014;**4**:4762

[17] Zhang X, Chen Q, Hu W, Zhang J. A DFT study of SF_6 decomposed gas adsorption on an anatase (101) surface. Applied Surface Science. 2013;**286**:47-53

[18] Zhang X, Chen Q, Hu W, Zhang J. Adsorptions of SO_2, SOF_2, and SO_2F_2 on Pt-modified anatase (1 0 1) surface: Sensing mechanism study. Applied Surface Science. 2015;**353**: 662-669

[19] Zhang X, Dai Z, Chen Q, Tang J. A DFT study of SO_2 and H_2S gas adsorption on Au-doped single-walled carbon nanotubes. Physica Scripta. 2014;**89**:065803

[20] Zhang X, Luo C, Tang J. Sensitivity characteristic analysis of adsorbent-mixed carbon nanotube sensors for the detection of SF_6 decomposition products under PD conditions. Sensors 2013;**13**:15209

[21] Zhang WL, Liu ZZ, Wang MJ. Yang XS. Research status and development trend of smart grid. Power System Technology. 2009;**33**:1-11

[22] Xue-Hao HU. Smart grid—A development trend of future power grid. Power System Technology. 2009;**33**:1-5

[23] Tang J, Ma S, Zhang M, Liu Z. Influence of microbubbles motion state on partial discharge in transformer oil. IEEE Transactions on Dielectrics and Electrical Insulation. 2015;**22**:2646-2652

[24] Tang J, Ma S, Zhang X, Zhang M. Investigation of partial discharge between moving charged metal particles and electrodes in insulating oil under flow state and AC condition. IEEE Transactions on Dielectrics and Electrical Insulation. 2016;**23**:1099-1105

[25] Kassi KS, Fofana I, Meghnefi F, Yeo Z. Impact of local overheating on conventional and hybrid insulations for power transformers. IEEE Transactions on Dielectrics and Electrical Insulation. 2015;**22**:2543-2553

[26] Jin WU, Kang-Jian XU. Probe into transformer oil dissolved gas three-ration method. Shaanxi Electric Power. 2011

[27] Kun XU, Zhou J, Qiushi RU. Zhou Z. Development and prospect of transformer oil dissolved gas On-line monitoring technology. High Voltage Engineering. 2005

[28] Aghaei J, Gholami A, Shayanfar HA, Dezhamkhooy A. Dissolved gas analysis of transformers using fuzzy logic approach. European Transactions on Electrical Power. 2010;**20**:630-638

[29] Singh J, Sood YR, Jarial RK. Condition monitoring of power transformers—Bibliography survey. IEEE Electrical Insulation Magazine. 2008;**24**:11-25

[30] Lin C-H, Chen J-L, Huang P-Z. Dissolved gases forecast to enhance oil-immersed transformer fault diagnosis with grey prediction-clustering analysis. Expert Systems. 2011;**28**:123-137

[31] Laisheng T, Guangning W, Jinlu S, Jun Z, Lijun Z. Oil-gas separation mechanism of polymer membranes applied to online transformer dissolved gases monitoring. In: Conference Record of the 2004 IEEE International Symposium on Electrical Insulation; September 19-22, 2004; IEEE; 2004. pp. 97-100

[32] Chatterjee A, Sarkar R, Roy NK, Kumbhakar P. Online monitoring of transformers using gas sensor fabricated by nanotechnology. International Transactions on Electrical Energy Systems. 2013;**23**:867-875

[33] Thang KF, Aggarwal RK, Esp DG, McGrail AJ. Statistical and neural network analysis of dissolved gases in power transformers. In: Eighth International Conference on Dielectric

Materials, Measurements and Applications; September 17-21, 2000; Edinburgh, UK; 2000; (IEE Conf. Publ. No. 473). pp. 324-329

[34] Ma GM, Li CR, Mu RD, Jiang J, Luo YT. Fiber bragg grating sensor for hydrogen detection in power transformers. IEEE Transactions on Dielectrics and Electrical Insulation. 2014;**21**:380-385

[35] Mao S, Lu G, Chen J. Nanocarbon-based gas sensors: Progress and challenges. Journal of Materials Chemistry A. 2014;**2**:5573-5579

[36] Kauffman DR. Star A. Carbon nanotube gas and vapor sensors. Angewandte Chemie International Edition. 2008;**47**:6550

[37] Su J, Cao L, Li L, Wei J, Li G, Yuan Y. Highly sensitive methane catalytic combustion micro-sensor based on mesoporous structure and nano-catalyst. Nanoscale. 2013;**5**:9720

[38] Yamazoe N. Toward innovations of gas sensor technology. Sensors and Actuators B Chemical. 2005;**108**:2-14

[39] Mizsei J. How can sensitive and selective semiconductor gas sensors be made? Sensors and Actuators B Chemical. 1995;**23**:173-176

[40] Jin W, Stewart G, Culshaw B, Murray S. Source-noise limitation of fiber-optic methane sensors. Applied Optics. 1995;**34**:2345-2349

[41] Agbor NE, Petty MC, Monkman AP, Cresswell JP. An optical gas sensor based on polyaniline Langmuir-Blodgett films. Sensors and Actuators B Chemical. 1997;**41**:137-141

[42] Gu Z, Xu Y, Gao K. Optical fiber long-period grating with solgel coating for gas sensor. Optics Letters. 2006;**31**:2405-2407

[43] Tierney MJ, Kim HOL. Electrochemical gas sensor with extremely fast response times. Analytical Chemistry. 1993;**65**:3435-3440

[44] Funazaki N, Kume S, Hemmi A, Ito S, Asano Y, Yamashita S. Development of catalytic electrochemical gas sensor for arsine. Sensors and Actuators B Chemical. 1993;**13**:466-469

[45] Rai P, Khan R, Raj S, Majhi SM, Park KK, Yu YT, Lee IH, Sekhar PK. Au@Cu_2O core-shell nanoparticles as chemiresistors for gas sensor applications: effect of potential barrier modulation on the sensing performance. Nanoscale. 2014;**6**:581-588

[46] Howe RT, Muller RS. Resonant-microbridge vapor sensor. IEEE Transactions on Electron Devices. 1986;**33**:499-506

[47] Qin LC, Zhao X, Hirahara K, Miyamoto Y, Ando Y, Iijima S. Materials science: The smallest carbon nanotube. Nature. 2000;**408**:50

[48] Kong J, Franklin NR, Zhou C, Chapline MG, Peng S, Cho K, Dai H. Nanotube molecular wires as chemical sensors. Science. 2000;**287**:622-625

[49] Kong J, Chapline MG, Dai H. ChemInform abstract: Functionalized carbon nanotubes for molecular hydrogen sensors. ChemInform. 2001;**13**:1384-1386

[50] Qi P, Vermesh O, Grecu M, Javey A, Wang Q, Dai H, Peng S, and, Cho KJ. Toward large arrays of multiplex functionalized carbon nanotube sensors for highly sensitive and selective molecular detection. Nano Letters. 2003;**3**:347-351

[51] Varghese OK, Kichambre PD, Gong D, Ong KG, Dickey EC, Grimes CA. Gas sensing characteristics of multi-wall carbon nanotubes. Sensors and Actuators B Chemical. 2001;**81**: 32-41

[52] Modi A, Koratkar N, Lass E, Wei B, Ajayan PM. Miniaturized gas ionization sensors using carbon nanotubes. ChemInform. 2003;**424**:171-174

[53] Robinson JT, Perkins FK, Snow ES, et al. Reduced graphene oxide molecular sensors. Nano Letters. 2008;**8**(10):3137

[54] Zhang Y, Liu J, Li X, Zhu C. The structure optimization of the carbon nanotube film cathode in the application of gas sensor. Sensors and Actuators A Physical. 2006;**128**:278-289

[55] Bondavalli P, Legagneux P, Pribat D. Carbon nanotubes based transistors as gas sensors: State of the art and critical review. Sensors and Actuators B Chemical. 2009;**140**:304-318

[56] Zhang Y, He J, Xiao P, Gong Y. Flow sensing characteristics of thin film based on multiwall carbon nanotubes. International Journal of Modern Physics B. 2012;**21**:3473-3476

[57] Goldoni A, Larciprete R, Petaccia AL, Lizzit S. Single-wall carbon nanotube interaction with gases: Sample contaminants and environmental monitoring. Journal of the American Chemical Society. 2003;**125**:11329-11333

[58] Li J, Lu Y, Qi Y, Cinke M, Jie H, Meyyappan M. Carbon nanotube sensors for gas and organic vapor fetection. Nano Letters. 2003;**3**:929-933

[59] Zhou Z, Gao X, Yan J, Song D. Doping effects of B and N on hydrogen adsorption in single-walled carbon nanotubes through density functional calculations. Carbon. 2006;**44**: 939-947

[60] Khare BN, Meyyappan M, Cassell AM, Cattien A, Nguyen V, Han J. Functionalization of carbon nanotubes using atomic hydrogen from a glow discharge. Nano Letters. 2002;**2**:73-77

[61] Cahill LS, Yao Z, Adronov A, Penner J, Moonoosawmy KR, Kruse AP, Goward GR. Polymer-Functionalized carbon nanotubes investigated by Solid-State nuclear magnetic resonance and scanning tunneling microscopy. Journal of Physical Chemistry B. 2004;**108**:11412-11418

[62] Chen RJ, Zhang Y, Wang D, Dai H. Noncovalent sidewall functionalization of single-walled carbon nanotubes for protein immobilization. Journal of the American Chemical Society. 2001;**123**:3838-3839

[63] Curran SA, Ajayan PM, Blau WJ, Carroll DL, Coleman JN, Dalton AB, Davey AP, Drury A, McCarthy B, Maier S, Strevens A. A composite from poly(m-phenylenevinylene-co-2,5-dioctoxy-p-phenylenevinylene) and carbon nanotubes: A novel material for molecular optoelectronics. Advanced Materials. 1998;1091-1109

[64] Hiura H, Ebbesen TW, Tanigaki K. Opening and purification of carbon nanotubes in high yields. Advanced Materials. 1995;7:275-276

[65] Holzinger M, Vostrowsky O, Hirsch A, Hennrich F, Kappes M, Weiss R, Jellen F. Sidewall functionalization of carbon nanotubes this work was supported by the European Union under the 5th Framework Research Training Network 1999, HPRNT 1999-00011 FUNCARS. Angewandte Chemie International Edition. 2001;**40**:4002-4005

[66] Liu J, Rinzler AG, Dai H, Hafner JH, Bradley RK, Boul PJ, Lu A, Iverson T, Shelimov K, Huffman CB. Fullerene pipes. Science. 1998;**280**:1253-1256

[67] Peng H, Gu Z, Yang J, Zimmerman JL, Willis PA, Bronikowski MJ, Smalley RE, Hauge RH, Margrave JL. Fluorotubes as cathodes in lithium electrochemical cells. Nano Letters. 2001; **1**:625-629

[68] Star A. Starched carbon nanotubes. Angewandte Chemie International Edition. 2002;**41**: 2508-2512

[69] Tsang SC, Chen YK, Harris PJF, Green MLH. A simple chemical method of opening and filling carbon nanotubes. Nature. 1994;**372**:159-162

[70] Zhao Q, Nardelli MB, Lu W, Bernholc J. Carbon nanotubes-metal cluster composites: A new road to chemical sensors. Nano Letters. 2005;**5**:847-851

[71] Peng S, Cho K. Ab Initio study of doped carbon nanotube sensors. Nano Letters. 2003;**3**: 513-517

[72] Ebbsen TW, Ajayan PM. Large-scale synthesis of carbon nanotubes. Nature. 1992;**358**(6383):220-222

[73] Journet C, Maser WK, Bernier P, et al. Large-scale production of single-walled carbon nanotubes by the electric-arc technique. Nature International Weekly Journal of Science. 1997;**388**(6644):756-758

[74] Mingliang S, Hongqiang W, Xinhai L, et al. Producing carbon nanotubes with arc discharge method. Journal of Xiangtan Mining Institute, 1999;**14**(1):54-57

[75] Ebbesen TW, Ajayan PM. Large-scale synthesis of carbon nanotubes. Nature. 1992;**358**: 220-222

[76] Thess A, Lee R, Nikolaev P, et al. Crystalline ropes of metallic carbon nanotubes. Science. 1996;**273**(5274):483

[77] Sen R, Govindaraj A, Rao CNR. Carbon nanotubes by the metallocene route. Chemical Physics Letters. 1997;**267**(3):276-280

[78] Chernozatonskii LA, Kosakovskaja ZJ, Fedorov EA, et al. New carbon tubelite-ordered film structure of multilayer nanotubes. Physics Letters A. 1995;**197**(1):40-46

[79] Yamamoto K, Koga Y, Fujiwara S, et al. New method of carbon nanotube growth by ion beam irradiation. Applied Physics Letters. 1996;**69**(27):4174-4175

[80] Lin X, Wang XK, Dravid VP, Chang RPH, Ketterson JB. Large scale synthesis of single-shell carbon nanotubes. Applied Physics Letters. 1994;**64**:181-183

[81] Richter H, Hernadi K, Caudano R, et al. Formation of nanotubes in low pressure hydrocarbon flames. Carbon. 1996;**34**(3):427-429

[82] Das Chowdhury K, Howard JB, Vandersande JB. Fullerenic nanostructures in flames. Journal of Materials Research. 1996;**11**(2):341-347

[83] Laplaze D, Bernier P, Maser WK, et al. Carbon nanotubes: The solar approach. Carbon. 1998;**36**(5):685-688

[84] Hsu WK, Terrones M, Hare JP, et al. Electrolytic formation of carbon nanostructures. Chemical Physics Letters. 1996;**262**(1):161-166

[85] Hui H, Wenkui Z, Chunan M, et al. Preparation of carbon nanotubes and nanowires by electrolysis in molten salts. Chinese Journal of Chemical Physics. 2003;**16**(2):131-134

[86] Stevens MG, Subramoney S, Foley HC. Spontaneous formation of carbon nanotubes and polyhedra from cesium and amorphous carbon. Chemical Physics Letters. 1998;**292**(3):352-356

[87] Chernozatonskii LA, Val'Chuk VP, Kiselev NA, et al. Synthesis and structure investigations of alloys with fullerene and nanotube inclusions. Carbon. 1997;**35**(6):749-753

[88] Yan Y, Wang WQ, Zhang LX. Dynamical behaviors of fluid-conveyed multi-walled carbon nanotubes. International Journal of Modern Physics B. 2009;**33**:1430-1440

[89] Kyotani T, Lifu Tsai A, Tomita A. Preparation of ultrafine carbon tubes in nanochannels of an anodic aluminum oxide film. Chemistry of Materials. 1996;**8**(8):2109-2113

[90] Matveev AT, Golberg D, Novikov VP, et al. Synthesis of carbon nanotubes below room temperature. Carbon. 2001;**39**(1):155-158

[91] Cantalini C, Valentini L, Armentano I, Kenny JM, Lozzi L, Santucci S. Carbon nanotubes as new materials for gas sensing applications. Journal of the European Ceramic Society. 2004;**24**:1405-1408

Chapter 6

Fabrication and Characterization of Metal-Loaded Mixed Metal Oxides Gas Sensors for the Detection of Hazardous Gases

Chang-Seop Lee and Yong Jae Kim

Additional information is available at the end of the chapter

http://dx.doi.org/10.5772/intechopen.68414

Abstract

This study concerns gas sensors that may protect individuals by detecting hazardous gases that may be generated in hot spaces (≥50°C) with residues of organic waste. We investigated the responses and selectivities of the sensors to different kinds of hazardous gases such as acetaldehyde, toluene and hydrogen sulfide. We also investigated operating temperatures and catalysts for the sensors. The thick film semiconductor sensors that detected some hazardous gases were prepared using nano-sized sensing material powders (SnO_2, WO_3, ZnO) that were prepared through sol-gel and precipitation methods. The nano-sized sensing materials were blended with various amounts of metal oxides (SnO_2, ZnO, WO_3) and coated with transition metals (Pt, Pd, Ru, Au, Ag, Cu and In). The metal oxide thick films were fabricated on an Al_2O_3 plate with a Ni-Cr heater and a Pt electrode through a screen-printing method. Morphologies, compositions, phases, surface areas and particle sizes of sensor compounds were examined by SEM, EDS, XRD and BET analysis. The investigated response to the various hazardous vapors was expressed as the value of Ra/Rg, where Ra and Rg are the resistance of the sensor material in the air and in hazardous gas, respectively.

Keywords: gas sensors, hazardous gas, metal oxide, screen-printing, sol-gel, thick film

1. Introduction

The semiconducting metal oxide gas sensors currently constitute one of the most investigated groups of gas sensors. Semiconducting metal oxide gas sensors are electrical conductivity sensors with active sensing layers that exhibit resistance changes when coming into contact with the hazardous gas under investigation. Metal oxide gas sensors are appropriate for a wide range of applications and for the detection of all reactive gases due to unique atomic-level chemistry and properties of each device.

Hydrocarbons, nitrogen oxides, oxidized hydrocarbons and sulfur oxides are common air contaminants in everyday life. Of these chemicals, hazardous gases produced from organic solvents used in industrial and laboratory processes are becoming an area of increasing social concern, since those vapors have a direct damage on the human body. Acute and chronic exposures to hazardous gases may cause humans to develop central nervous system disorders, reproductive system diseases and leukemia [1–3].

Formaldehyde and acetaldehyde are known to be two of the carbonyl compounds most widely disseminated in ambient air. Aldehydes products produced by combustion and oxidation can interact directly with ozone. The products are major components of smog, and further photochemical reactions may enhance global warming by destroying ozone layer in the stratosphere. Aldehydes commonly present in the gaseous state at 25°C and stay in the air for a long period of time. Acetaldehyde is the main component of hazardous vapor produced in newly built houses and can pollute the air both outdoors and indoors. Humans exposed to aldehydes irritations to their respiratory systems such as their throats, noses and bronchi. In addition, acetaldehyde may affect the central nervous system of an individual via anesthesia. Extreme physical reactions such as paralysis, respiratory disorders, and even the onset of a comatose state are among the events that can occur in extreme cases of acetaldehyde exposure [4, 5]. Therefore, for both environmental and for health reasons, it is very important to detect the presence of aldehydes.

Hydrogen sulfide, a byproduct of the active decomposition of organic matter, is very hazardous gas. For instance, many people died in the underground sewage systems of Paris during the French Revolution. According to the U.S. government report, H_2S is a watchful gas for instant death in the working place. Because H_2S gas is rapidly oxidized into sulfides that can easily be eliminated via the kidneys, the gas is not known to have a cumulative toxic effect. Inhalation of lethal concentrations (1000–2000 ppm) of hydrogen sulfide may cause olfactory fatigue or paralysis and ultimately the cessation of breath. Other forms of respiratory distress known to occur after inhalation of hydrogen sulfide (500–1000 ppm) are dose-dependent deficits such as hyperpnea (via carotid body stimulation) and sleep apnea. Low levels of hydrogen sulfide (50–500 ppm) can act as primary irritants to the eyes and the respiratory system. Continuous exposure to 250–600 ppm of the hazardous gas may result in pulmonary edema.

Toluene is a colorless volatile liquid with a unique smell. Safe handling of this highly volatile compound includes storage in a perfectly sealed container. If toluene is handled in an industrial workplace or in a laboratory without wearing special protective equipment, a worker or researcher may inhale 100% of the toluene. Prolonged inhalation exposure to toluene, even at low concentrations, can lead to deterioration of lung function and eyesight. Additional effects include a headache, paralysis of the hands/feet, cancer, leukemia, mental derangement, malignant lymphoma and depression of central nervous system.

In this chapter, we describe the development of catalyst-loaded metal oxide gas sensors to detect gases that posed hazards to the human body, that is, acetaldehyde gas, hydrogen sulfide and toluene vapors.

2. Experimental details

2.1. Preparation of sensor materials

One of the disadvantages of gas sensors is the costly requirement for more electricity to maintain a high operating temperature. To mitigate this disadvantage, there is a need to increase the electronic conductivity of the sensor materials in order to decrease the operating temperature. One way to improve the electronic conductivity is to add conducting materials to sensor materials.

For the acetaldehyde gas sensor, we synthesized WO_3 through a sol-gel process and used it as a sensing material to detect acetaldehyde gas. Metal catalysts were added through the impregnation method [4–6], to enhance the sensitivity and selectivity that are functional prerequisites of the gas sensor. Transition metals (Pd, Ru, Pt and In) were weighed to give weight ratios of 1–5 wt%, and 1 ml of hydrochloric acid (HCl) diluted with distilled water was added to completely dissolve the metal catalysts in a beaker. Heating and stirring of appropriate suspensions with magnetic stirrers yielded WO_3 or SnO_2/WO_3 powders impregnated with metal catalysts. Then, we dried each solution for 12 h at 100°C and calcined it for 2 h at 500°C to get powder. Finally, we obtained a fine powder of the sensing material by crushing the crude material.

We added metal catalysts to WO_3/SnO_2 powder to prepare the sensing material by using the same methods described above in the case of the hydrogen sulfide sensor. Additional techniques included co-precipitation, impregnation method and alleged fabrication [6]. We put various metals of Pd, Ru, Ag, Au and In as catalysts in the flask with the ratios of 0.5–5 wt% and dissolved the metal catalysts by adding 1 ml of concentrated HCl solution. We then added SnO_2/WO_3 powder to impregnate the metal catalysts. These materials were slowly heated while stirring with a magnetic stirrer for an effective loading of metal catalysts to SnO_2/WO_3. The as-prepared sample was then dried for 24 h at 110°C and calcinated for 2 h at 600°C. Finally, the sample was pulverized to make each one of the sensor materials.

In the case of the toluene sensor, we produced a sensor material by adding carbon black using an impregnation method. An impregnation method was also used for the addition of the transition metal. Carbon black was measured to be at 1–20 wt.% of the main material. Then, it was put into a beaker and 2 ml of HCl was added. After transition metal and carbon black are completely dissolved, distilled water was added to make a homogeneous solution. Next, the transition metal and carbon black were impregnated by adding SnO_2/ZnO powder. Then, it was gradually heated while stirring with a magnetic stirrer to produce SnO_2/ZnO/catalyst powder with carbon black deposit. The powder was dried for 24 h at 110°C and went through calcination for 5 h at 600°C. The desired material was obtained by crushing the obtained powder [7–11].

2.2. Fabrication of thick film

A commercial product which has Pt electrode at 0.5-mm interval and a Ni-Cr heater in a 10 mm × 8 mm × 0.65 mm Al_2O_3 base plate was used as the supporting material for the sensor. The plate went through a washing process using an acetone and heat treatment process at 300°C for 1 h in order to remove contaminants before fabrication of the thick film. Next,

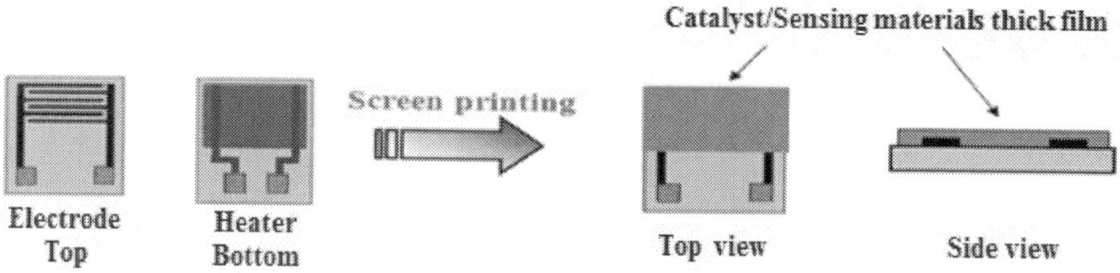

Figure 1. Structure of thick film sensor.

the sensor material paste was coated by a screen-printing method at a thickness of 20–60 μm. **Figure 1** shows diagrams before and after coating of the thick films [12].

Regarding thick film formation on each sensor material, a paste with proper viscosity was screen-printed on an alumina base plate with electrode formation achieved by adding ethylene glycol as a binder at 15–30 wt.% of powder. The resulting thick film element was left at room temperature for 12 h. Then, it was dried for 24 h in an 110°C oven. After 2 h calcination at 600°C, the final product was obtained.

2.3. Analysis of sensitivity characteristics

After we had installed the gas sensor at a distance of 50 mm from the bottom of a 10 L (250 mm x 200 mm x 200 mm) chamber, the sensitivity of the gas sensor was measured with the temperature range from 200 to 400°C. A stabilization procedure was performed for 12 h in the

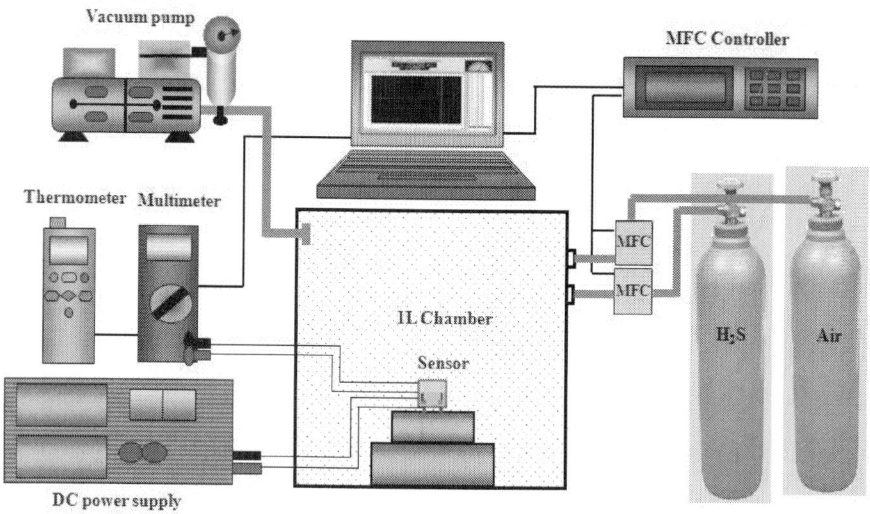

Figure 2. Apparatus used for gas-sensing experiments.

http://dx.doi.org/10.5772/intechopen.68414

same temperature range to remove the electrons from the system. Then, the target gas had been dispersed using a fan in the chamber. When the equilibrium concentration was reached, the resistance was measured with an electrometer. The sensor sensitivity was expressed as the value of Ra/Rg, where Ra and Rg are the resistance of the sensor material in the air and in target gas, respectively. The experimental apparatus is shown in **Figure 2**.

3. Results and discussion

3.1. Analysis results of sensing materials used for the gas sensor

3.1.1. The acetaldehyde gas sensor

In this study, we prepared sensing materials for a gas-detecting sensor using various composition ratios of WO_3 powder employed as the main sensing element to metal oxide (SnO_2) and metal catalysts. XRD, SEM/EDS and BET methods were employed to characterize the crystallinity and morphology, phase, composition and specific surface areas. The XRD patterns of WO_3 at various calcination temperatures and those when SnO_2 and metal catalysts were added to WO_3 were shown in **Figures 3** and **4**, respectively. In **Figure 3**, the peak intensity of WO_3, which is a characteristic of crystallization, rises with increasing calcination temperatures. This means that the crystallinity is enhanced at higher calcination temperature. By comparing the data we obtained with those of JCPDS as reference data, we concluded that the experimental data were agreed with an orthorhombic structure, which is one of the possible structures of WO_3 [13]. In **Figure 4**, new peaks started to appear at characteristic 2θ values as metal catalysts and metal oxides were added.

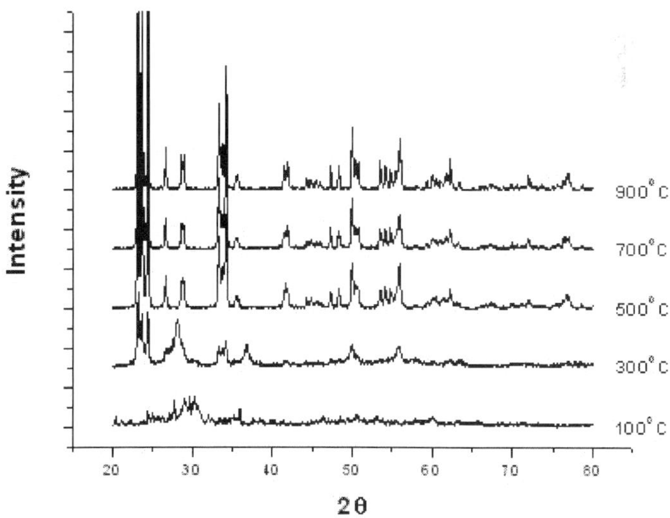

Figure 3. X-ray diffraction patterns of WO_3 powder with temperature.

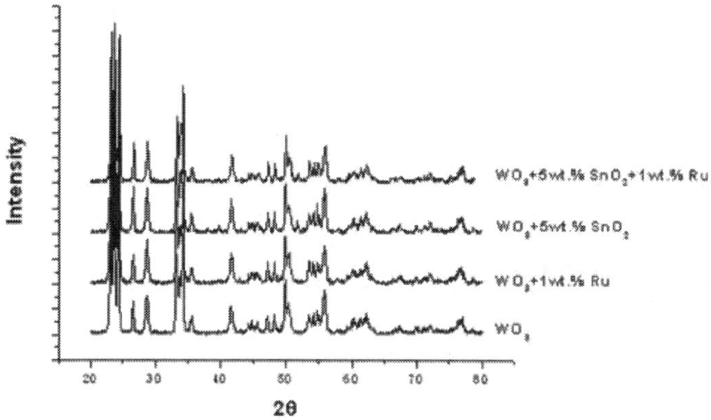

Figure 4. X-ray diffraction patterns of the various sensor materials based on catalyst/SnO_2/WO_3.

Figure 5. SEM image of WO3 powders calcined for 2 h at (a) 300°C, (b) 500°C, (c) 700°C and (d) 900°C.

Figure 5 exhibits SEM images of the WO_3 thick film at various calcination temperatures. These SEM images show that the particle size of the WO_3 thick film increases with rising calcination temperatures.

Figure 6 represents SEM images and EDS results of each sensing material composed of WO_3 containing metal oxides and metal catalysts. **Figure 6a** shows SEM photos and EDS results for sensor material of WO_3 only. **Figure 6(b)**, **(c)**, and **(d)** shows SEM photos and EDS data for sensor materials of 5 wt% SnO_2/WO_3, 1 wt% Ru/WO_3, and 1 wt% Ru/5 wt% SnO_2/WO_3, respectively. The SEM photos obviously show the surface morphology of each sensor material. The EDS data indicate the appearance of new peaks of the component in each sensor material.

3.1.2. The H_2S gas sensor

A sensing material for gas detection was fabricated with various mixing ratios of the main material SnO_2, metal oxide and catalyst. The techniques of SEM, EDS, XRD and BET were employed to analyze surface morphology, component, the degree of crystallinity, phase and surface area of each fabricated sensing material. **Figure 7** shows the XRD patterns of the SnO_2 synthesized by sol-gel method and SnO_2 mixed with catalyst and metal oxide.

By a comparison of the data of SnO_2 we obtained in this experiment with those in JCPDS, we presume the structure to be rutile form which is shown in **Figure 7**. As for the sample of SnO_2 mixed with 10 wt% WO_3, a new peak was appeared around 2θ. Referring to the data in JCPDS, the data of WO_3 tell us that the structure is orthorhombic among the three possible structures. Based on the interpretation of peaks of SnO_2, we can say that there was no specific phase change. Any new 2θ value was not also observed in the diffraction pattern for the sample of SnO_2 mixed with 10 wt% WO_3/1 wt% Ru. This implies that 1 wt% of Ru was not sufficient to produce a crystalline peak of Ru.

Table 1 shows the measured results of specific surface areas of SnO_2 in which pure SnO_2 was mixed with 10 wt% WO_3 and various metal catalysts. Compared to pure SnO_2, SnO_2 with added 10 wt% WO_3 and various metal catalysts showed larger BET surface areas, pore diameters and pore volumes. This means that SnO_2 added with WO_3 and metal catalyst hindered the particle growth, which results in increasing the specific surface area.

In particular, the SnO_2 powder (in which 10 wt% WO_3 was mixed with 1 wt% Ru) showed the highest BET specific surface area, pore diameter and total pore volume. The sensor fabricated with this material showed good functional properties, including response, reaction, recovery and selectivity.

SEM photos and EDS data for SnO_2 powder added with WO_3 and various metals were shown in **Figure 8**. In SEM photos, the surface morphology of each sensing material could be observed, and the peak of the component element for each sensing material was verified by the EDS data.

3.1.3. The toluene gas sensor

This study produced sensor material for gas detection by varying the mixing ratios of SnO_2 powder as main material, ZnO, transition metal catalyst and carbon black. The crystal property of each produced sensor material was analyzed by using XRD measurements. **Figure 9** shows each XRD pattern of various sensor materials such as SnO_2 synthesized by the precipitation method, SnO_2/ZnO which is SnO_2 with 20 wt.% ZnO, SnO_2/ZnO/Cu which is SnO_2/ZnO added of 1 wt.% Cu and SnO_2/ZnO/Cu/carbon black which is SnO_2/ZnO/Cu added of 5 wt.% carbon black. When the experimental data were compared to various values of SnO_2 and ZnO in JCPDS, it was confirmed that the peaks were characteristic of the crystal structure held by SnO_2 and ZnO. Meanwhile, it was not possible to confirm the peak of Cu and carbon black. The XRD

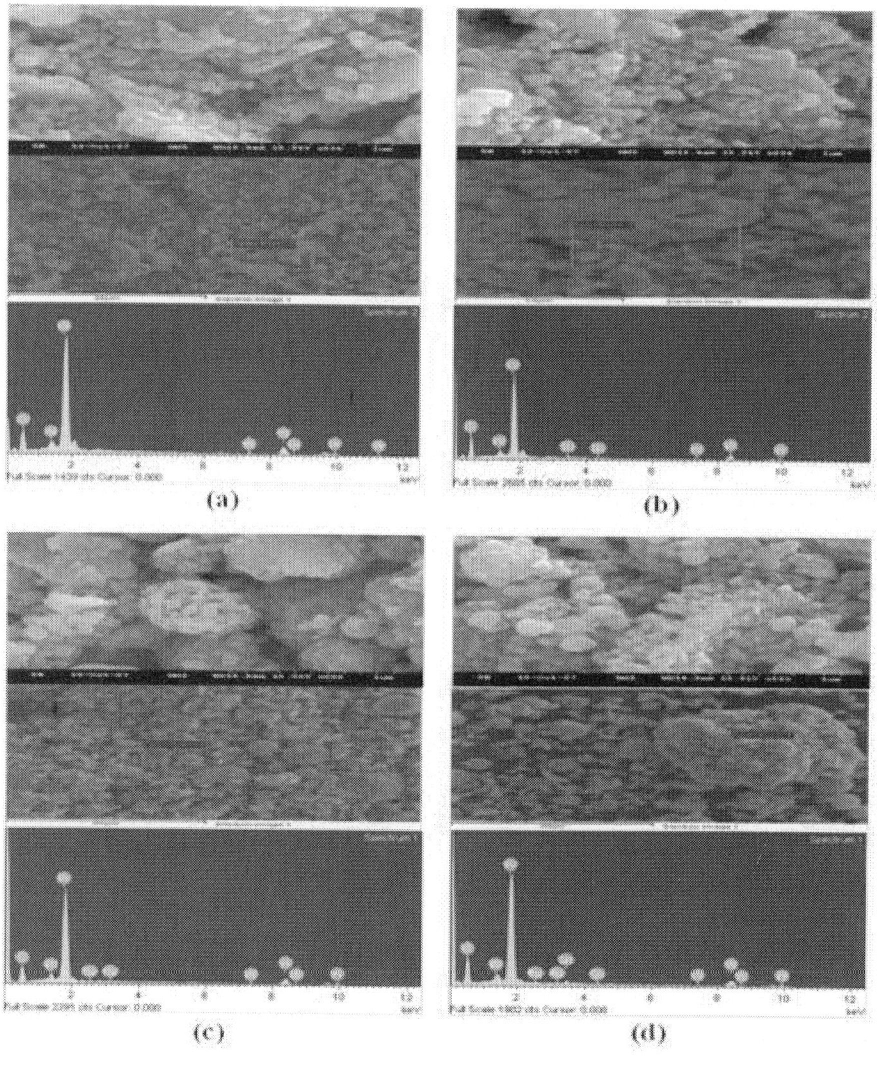

Figure 6. SEM images and EDS results of sensing materials. Upper: SEM (×50 K), middle: SEM (5 K) and lower: EDS (×5 K).

peak for Cu was not available because only small amount of Cu at 1 wt.% was added. The XRD peak for carbon black was not available because carbon black is a weak crystalline material.

Figure 10 shows SEM images taken at 100,000 magnification showing ratios of SnO_2, SnO_2/20%ZnO, SnO_2/20%ZnO/1%Cu and SnO_2/20%ZnO/1%Cu/5% carbon black, respectively.

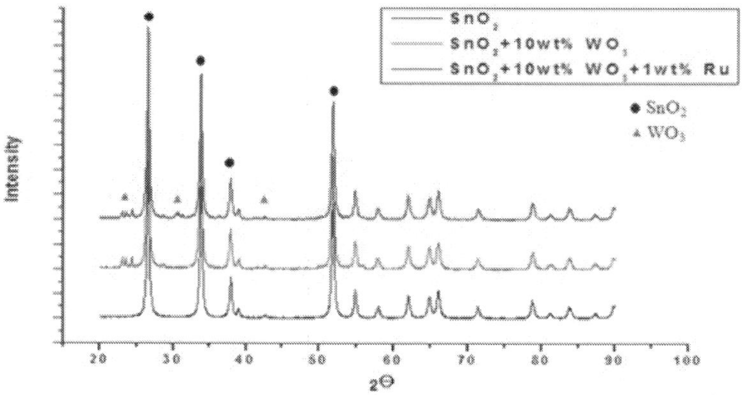

Figure 7. XRD patterns of catalyst/WO_3/SnO_2 powder.

Samples	BET surface area (m^2/g)	Pore diameter (nm)	Total pore volume (cm^3/g)
SnO_2	7.16	12.3	0.1218
1 wt% Ru-10 wt% WO_3/SnO_2	13.06	31.6	0.1398
1 wt% Pd-10 wt% WO_3/SnO_2	10.73	3.795	0.09435
1 wt% Au-10 wt% WO_3/SnO_2	10.64	21.01	0.09975
1 wt% In-10 wt% WO_3/SnO_2	11.53	26.92	0.1202
1 wt% Ag-10 wt% WO_3/SnO_2	11.39	22.34	0.09927

Table 1. Surface area (m^2/g), pore diameter (nm), pore volume (cm^3/g) of catalysts/WO_3/SnO_2.

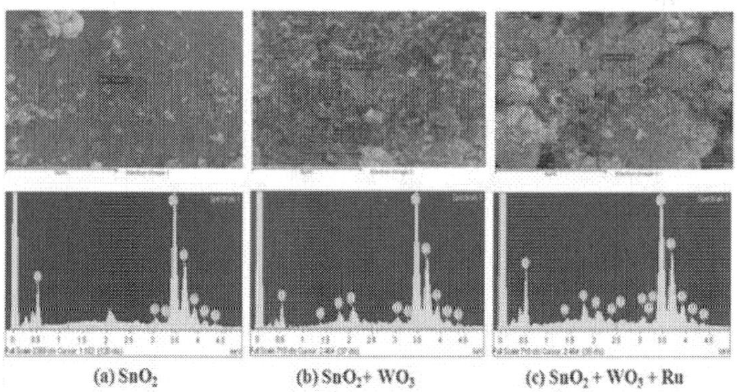

Figure 8. SEM images and EDS results of gas-sensing materials. (a) SnO_2, (b) 10 wt% WO_3/SnO_2 and (c) 1 wt% Ru/10 wt% WO_3/SnO_2.

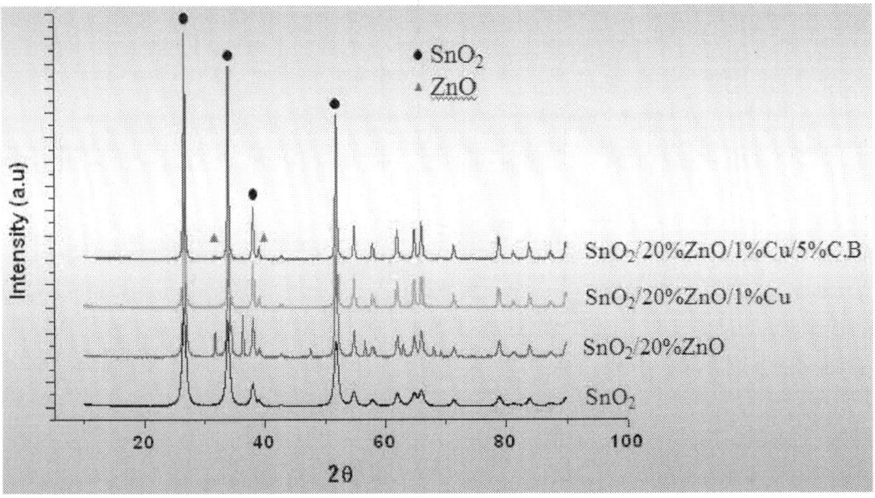

Figure 9. XRD patterns of the various sensor materials.

Figure 10. SEM images of the various sensor materials.

From these SEM images, it could be seen that the sensor material made of SnO_2/20%ZnO/1%Cu/5% carbon black had uniform particle sizes and crystallinities compared to other sensor materials.

Figure 11 shows the SEM images and EDS measurements for each sensor material composed of SnO_2, SnO_2/20%ZnO, SnO_2/20%ZnO/1%Cu and SnO_2/20%ZnO/1%Cu/5% carbon black, respectively. It was possible to observe the surface state of each sensor material through studying the SEM images shown in **Figure 11**. It was also possible to confirm the peaks of component elements existing in each sensor material from the EDS results. The existence of Sn and O was confirmed in **Figure 11(a)**; the existence of Sn, O and Zn was confirmed in **Figure 11(b)**; the existence of Sn, O, Zn and Cu was confirmed in **Figure 11(c)** and the existence of Sn, O, Zn, Cu and C was confirmed in **Figure 11(d)**.

3.2. Response characteristics to acetaldehyde gas

Figure 12 shows the sensing characteristics of WO_3 thick film sensor to acetaldehyde (100 ppm) at the various operating and calcination temperatures. In particular, WO_3 calcined at 500°C is shown to give the best sensitivity at 350°C. The sensing characteristics of the sensor after addition of SnO_2 and metal catalyst to acetaldehyde gas (100 ppm) at various operating temperatures were shown in **Figure 13**. While the highest sensitivity was reached at 350°C in the case of WO_3 alone, this sensitivity peak occurred at 300°C in the presence of additional constituents, that is, SnO_2 and Ru. This can be understood as follows: at a given temperature, the activation energy changes greatly depending on the amount of SnO_2 and the kind of metal catalyst added. Because the sensitivity of a sensor degrades with the temperature higher than 300°C, the experiments were carried out at a fixed operation temperature of 300°C.

Metal catalysts were added to the sensor material in order to improve the sensitivity and selectivity of the gas sensor to acetaldehyde gas. Mixing a metal catalyst with a metal oxide

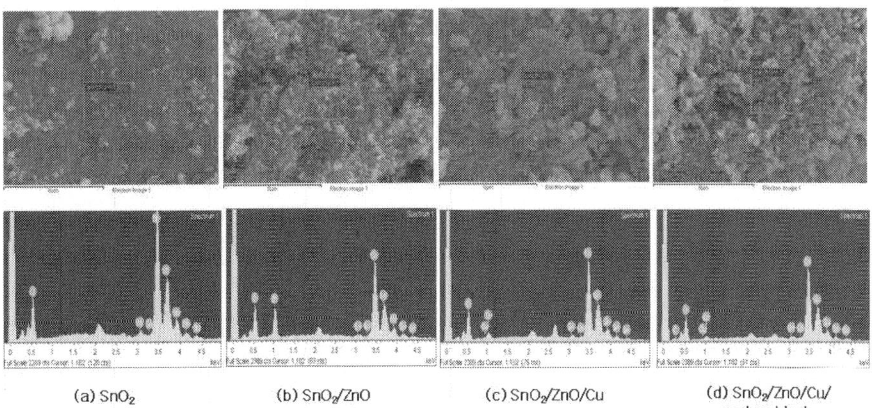

Figure 11. SEM images and EDS results of the various sensor materials.

induces an increase in the adsorption species caused by the activation of electrons. This affects the adsorption-desorption process, which results in an enhancement of electrical conductivity. To improve the effect of this catalytic reaction, the contacting area of metal catalysts with the sensor needs to be increased. Such a move has been found to greatly affect sensitivity, selectivity and operational temperature in terms of sensing the amount of gas [14].

Figure 14 depicts the sensing characteristics of sensor materials to acetaldehyde gas with various metal catalysts. The metal catalysts were added to 5 wt% SnO_2/WO_3, which was used as main sensor material. As shown in **Figure 14**, the sensor material containing 1 wt% of Ru gives better sensitivity than those of other catalysts. This explains that 5 wt% SnO_2/WO_3 with a metal catalyst of 1 wt% Ru shows better selectivity in oxidizing acetaldehyde gas

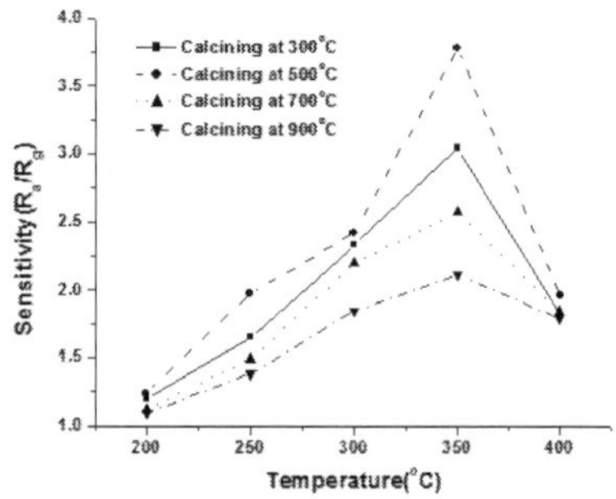

Figure 12. The sensitivity of WO_3 thick films with various calcination temperatures for acetaldehyde gas (100 ppm) with different operating temperatures.

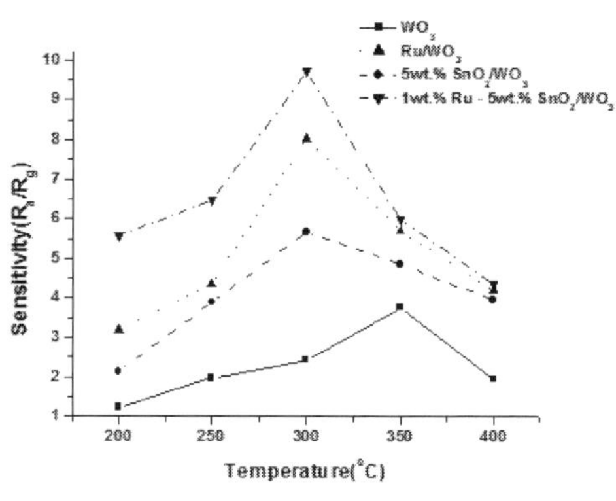

Figure 13. The sensitivity of the gas sensor for acetaldehyde gas (100 ppm) at various temperatures.

http://dx.doi.org/10.5772/intechopen.68414

than any other sensing materials. The reason we used 1 wt% concentration is that it gave the best sensitivity, that is, in the range of 0–1 wt%, the sensitivity increases with elevated Ru concentrations. On the other hand, if the concentration exceeds 1 wt%, there is a decrease in sensitivity. In this case, the Ru particles were not able to spread homogenously on the surface of the thick film due to the coagulation.

Figure 15 shows the results of the selectivity study of the sensing materials giving with the best selectivities toward acetaldehyde gas compared to various other volatile organic compounds. The sensitivity measurement of 1 wt% Ru/5 wt% SnO_2/WO_3 gas sensor which showed the best sensitivity to acetaldehyde gas was performed within a concentration range of 100 ppm. The result showed that the sensitivity to acetaldehyde gas was much higher than other gases. It means that the sensing material of 1 wt% Ru/5 wt% SnO_2/WO_3 works better for acetaldehyde gas in decomposing and activating reaction than for other VOC gases.

3.3. Sensing characteristics to H_2S gas

The sensing response of WO_3/SnO_2 was investigated for different type of sensor materials where the amounts of metal catalyst added were varied. The kinds of catalyst were selected to give a good response and selectivity to H_2S gas. In general, metal catalyst mixed with metal oxide helps to transfer electrons smoothly and increases the number of absorbed species in the absorption process, which lead to the change of electric conductivity on the sensor material. It has also been known that increasing the contact area affects the catalyst to enhance the response, selectivity and operating temperature of the sensor [15].

Figure 16 shows the response of hydrogen sulfide across different metal catalysts. Various metal catalysts were added to the basic sensor material composed of 10 wt% WO_3/SnO_2. As shown in **Figure 16**, the sensor material with added 1 wt% Ru compared to other metal catalysts, and the operating temperature was 200°C, which represented a reduction of 50°C. This meant that 1 wt% Ru/10 wt% WO_3/SnO_2 sensor material oxidized hydrogen sulfide better than other sensor materials and also showed better selectivity.

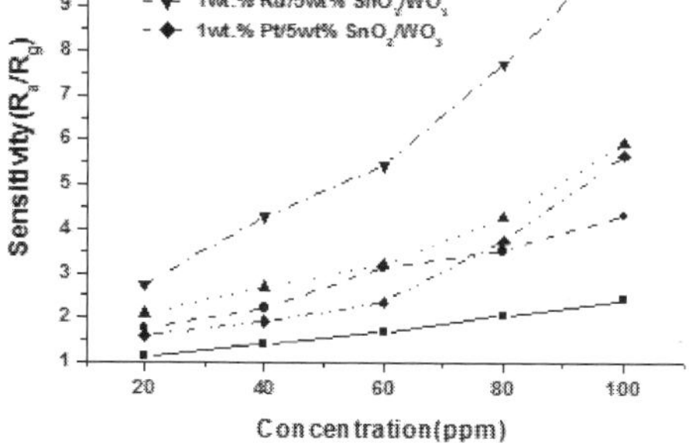

Figure 14. Sensitivity of catalyst/5 wt% SnO_2/WO_3 to various concentration of acetaldehyde gas at 300°C.

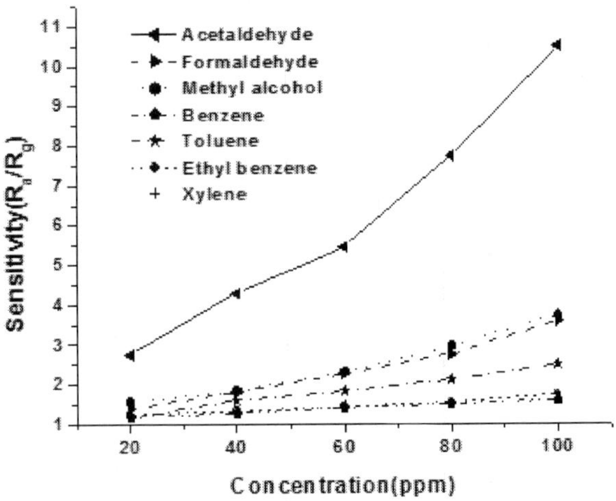

Figure 15. The sensitivity of 1 wt% Ru/5 wt% SnO_2/WO_3 to various gases (100 ppm) at 300°C.

Figure 17 shows the sensing response of H_2S gas for the samples of various amount of metal catalysts added to find the optimal condition. For the main sample of 10 wt% WO_3/SnO_2, the sensing response was examined by varying the amount of Ru catalyst added in the range of 0.5–5.0 wt%, and it was found to be the best for the case of 1 wt% Ru addition. When the quantity of catalyst was below 1 wt%, it is not sufficient to activate the reaction. While the quantity of catalyst was above 1 wt%, the Ru particles aggregated itself and changed the characteristics of sensor material which did not further improve responses.

Figure 18 shows the response comparing the changes in different concentration levels of 100 ppm of hydrogen sulfide at 200°C when SnO_2 was mixed with 10 wt% WO_3 + 1% Ru. Finally, the sensor material of 1 wt% Ru/10 wt% WO_3/SnO_2 showed the best response characteristics.

Figure 19 presents the selectivity results of various gases based on the sensor material showing the best sensitivity for H_2S gas. The sensor prepared with the material of 1 wt% Ru/10 wt% WO_3/SnO_2 showed the best sensitivity and was tested its sensitivities to various gases at an operating temperature of 200°C and the concentration range of 20–100 ppm. As a result, the sensor showed better response to H_2S gas than to other gases. Therefore, it is known that 1 wt% Ru/10 wt% WO_3/SnO_2 adsorbs H_2S gas well and enhances the catalytic reaction to improve the selectivity.

3.4. Response characteristics to toluene gas

Carbon black is a material made by covalent bonding of carbon. It is known that carbon black has good electric conductivity and good thermal conductivity. There is existing literature

http://dx.doi.org/10.5772/intechopen.68414

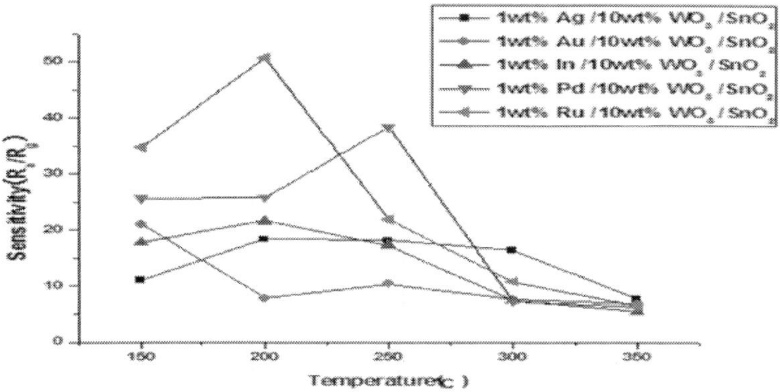

Figure 16. Response with catalysts/10 wt% WO_3/SnO_2 to the hydrogen sulfide gas (100 ppm) at different operating temperatures.

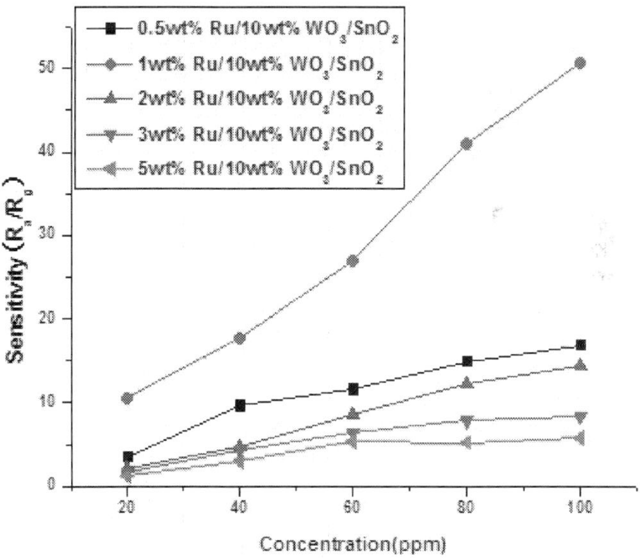

Figure 17. Sensitivity with the catalyst/WO_3/SnO_2 to the various concentrations of hydrogen sulfide gas at 200°C.

study that reported the addition of carbon black on sensor material to decrease sensor operation temperature [16–18]. In this study, carbon black is added to sensor material to decrease operation temperature and increase the sensitivity of the sensor. **Figure 20** is the sensitivity on 100 ppm toluene gas depends on carbon black added quantity. As seen in the figure, the

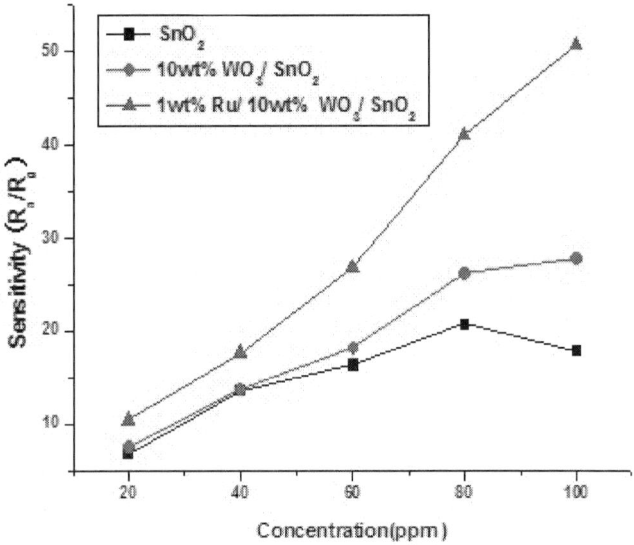

Figure 18. The sensitivity of 1 wt% Ru/10 wt% WO_3/SnO_2 from 20 ppm to 100 ppm of hydrogen sulfide gas at 200°C.

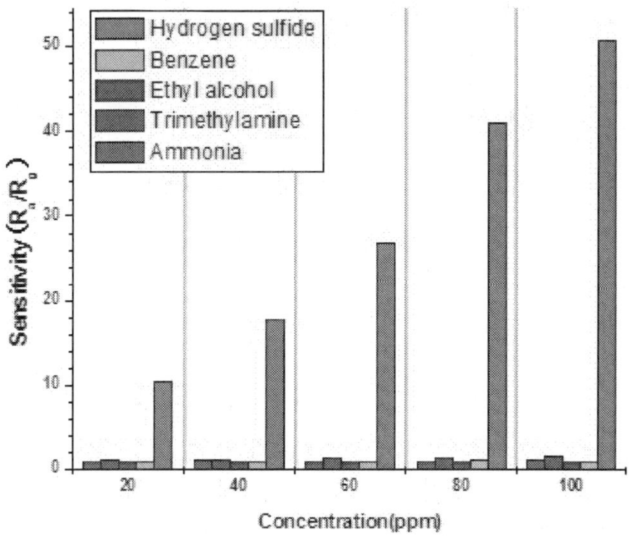

Figure 19. Sensitivity of 1 wt% Ru/10 wt% WO_3/SnO_2 to the various concentrations of various gases at 200°C.

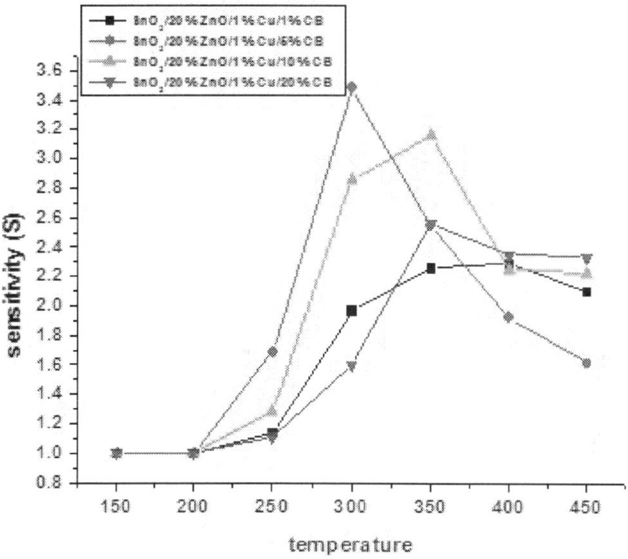

Figure 20. Sensitivity of sensors with various contents of carbon black to the toluene gas (100 ppm) at various temperatures.

sensitivity of sensor fabricated with material added of 5 wt.% carbon black of SnO_2 has better sensitivity and 50°C lower operation temperature compared to sensors fabricated with material added of different wt% of carbon black. The reason is believed that the sensor sensitivity and operation temperature decreased by the increase in heat dispersion in sensor material when carbon black was added.

Figure 21 is the sensitivity measurement on 100 ppm toluene gas using a sensor fabricated with different components such as SnO_2, SnO_2/20%ZnO, SnO_2/20%ZnO/1%Cu and SnO_2/20%ZnO/1%Cu/5%carbon black. As seen in the figure, the sensor fabricated with SnO_2/20%ZnO/1%Cu/5%carbon black has better sensitivity and lower operation temperature compared to sensors produced with other three materials.

Figure 22 is the sensitivity measurement on 1–20 ppm low concentration toluene gas using a sensor made of SnO_2/20%ZnO/1%Cu/5%carbon black, which showed the best sensitivity in this study. This sensor showed a linear increase in sensitivity at a concentration of 20 ppm or less in general.

Figure 23 is the selectivity on various volatile gases of the sensor fabricated with SnO_2/20%ZnO/1%Cu/5%carbon black, which showed the best sensitivity on toluene gas in this study. Sensitivity was measured on various gases at 300°C operation temperature and 10–100 ppm concentration range. The result suggested that the sensor had the best sensitivity on toluene gas compared to the sensitivity to other gases. There was some interference effect

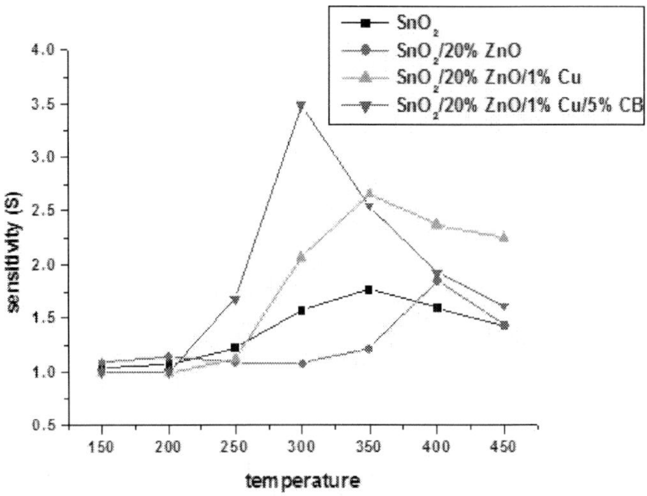

Figure 21. Sensitivity of the SnO_2/20%ZnO/1%Cu/5% carbon black sensor for toluene gas (100 ppm) to the various temperatures.

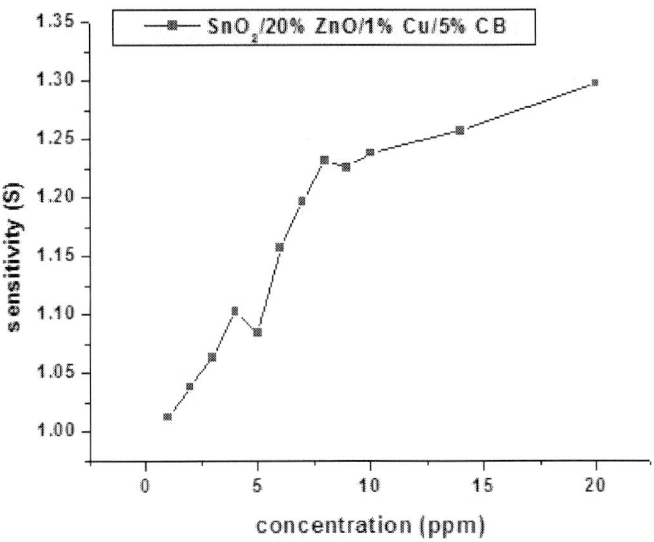

Figure 22. Sensitivity of SnO_2/20%ZnO/1%Cu/5%carbon black sensor from 1 to 20 ppm of toluene gas at 300°C.

http://dx.doi.org/10.5772/intechopen.68414

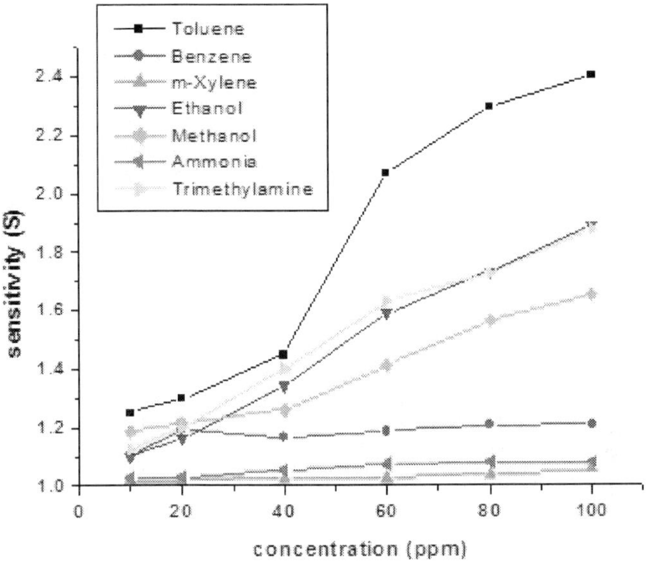

Figure 23. Sensitivity of SnO_2/20%ZnO/1%Cu/5%carbon black to the various gases at 300°C.

from ethanol and trimethylamine gas; however, interference effect from other volatile gases was not so significant. A sensor fabricated with SnO_2/20%ZnO/1%Cu/5%carbon black has better selectivity on toluene gas, because it decomposes more toluene gas and activates more reaction compared to other volatile gas.

4. Conclusion

Semiconductor metal oxide gas sensors can be used for diverse applications, ranging from equipment to monitor environmental and occupational safety to facilitating quality assurance through novel measurement. The nature of the gas-sensitive material and the concentration of the target gas (usually a few ppb–ppm) will determine the measuring range and limitations of the device.

In this study, the metal oxide gas sensor was fabricated with SnO_2 and WO_3 as the basic substances added to the metal oxide (SnO_2, WO_3, and ZnO) and various metal catalysts to detect the toxic gas. By changing the added amount and types of metal oxide and catalysts, this study investigated the sensitivity, operating temperature and recovery properties for the toxic gas.

(1) The sensing characteristics of the fabricated gas sensors added with catalysts and metal oxides to SnO_2 tend to rely on the operating temperature, calcination temperature and the kind and composition of the catalyst and metal oxide.

(2) The optimal conditions for the fabrication of the sensor material for detecting acetaldehyde gas were known to be 300°C as an operation temperature, 500°C as a calcination temperature, 20 μm as a film thickness and ethylene glycol as a binder. The best sensitivity, selectivity and response-recovery characteristics were achieved when 1 wt% Ru/5 wt% SnO_2 was added to the WO_3 sensing material.

(3) The sensing characteristics of hydrogen sulfide gas sensor made by adding catalyst and metal oxide to SnO_2 tend to rely on the operating temperature, calcination temperature and kind and composition of catalyst and metal oxide. The ideal condition for the sensor material to detect H_2S gas is temperatures of 200°C for operation and 600°C for calcination, and the sample of SnO_2 added with 10 wt% WO_3 and 1 wt% Ru.

(4) The optimum conditions for the toluene gas detection sensor are an operating temperature of 300°C with added metal catalyst Cu and carbon black. The best sensitivity and highest selectivity were achieved with 20% ZnO, 1% Cu and 5% carbon black, in terms of wt% to SnO_2 added to SnO_2 by impregnation.

Acknowledgements

This research was financially supported by the Ministry of Education (MOE) and National Research Foundation of Korea (NRF) through the Human Resource Training Project for Regional Innovation (no. 2015H1C1A1035858).

Author details

Chang-Seop Lee* and Yong Jae Kim

*Address all correspondence to: surfkm@kmu.ac.kr

Department of Chemistry, Keimyung University, Daegu, South Korea

References

[1] Wark K, Warner CF. Air Pollution: Its Origin and Control. New York: Haper Collins; 1981

[2] Noordally E, Richmond JR, Tahir SF. Destruction of volatile organic compounds by catalytic oxidation. Catalysis Today. 1993;**17**(1–2):359–366

[3] Hodgson AT, Faulkner D, Sullivan DP, DiBartolomeo DL, Russell ML, Fisk WJ. Effect of outside air ventilation rate on volatile organic compound concentrations in a call center. Atmospheric Environment. 2003;**37**(39–40):5517–5527

[4] Min BK, Choi SD. Development of methane gas sensor by various powder preparation methods. Journal of the Korean Ceramic Society. 1999;**5**(2):125–130

[5] Lim CB, Oh SJ. Microstructure evolution and gas sensitivities of Pd-doped SnO_2-based sensor prepared by three different catalyst-addition processes. Sensors and Actuators B: Chemical. 1996;**30**(3):223–231

[6] Yoo DJ, Tamaki J, Miura N, Yamazoe N, Park SJ. Effect of film thickness on gas sensing behavior of thin-film-type gas sensor. Korean Journal of Materials Research. 1996;**6**(7):716–722

[7] Acosta DR, Zironi EP, Montoya E, Estrada W. About the structural, optical and electrical properties of SnO_2 films produced by spray pyrolysis from solutions with low and high contents of fluorine. Thin Solid Films. 1996;**288**(1–2):1–7

[8] Serrini P, Briois V, Horrillo MC, Travers A, Manes L. Chemical composition and crystalline structure of SnO_2 thin films used as gas sensors. Thin solid Films. 1997;**304**(1–2):113–122

[9] Qi Q, Zhang T, Liu L, Zheng X. Improved NH_3, C_2H_5OH, and CH_3COCH_3 sensing properties of SnO_2 nanofibers by adding block copolymer P123. Sensors and Actuators B. 2009;**141**(1):174–178

[10] Zeng Y, Zhang T, Wang L, Kang M, Fan H, Wang R, He Y. Enhanced toluene sensing characteristics of TiO_2-doped flowerlike ZnO nanostructures. Sensors and Actuators B. 2009;**140**(1):73–78

[11] Song X, Zhang D, Fan M. A novel toluene sensor based on ZnO–SnO_2 nanofiber web. Applied Surface Science. 2009;**255**(16):7343–7347

[12] Kawahara A, Katsuki H, Egashira M. Fabrication of semiconductor oxide thick films by slide-off transfer printing and their NO_2-sensing properties. Sensors and Actuators B. 1998;49(3):273–278

[13] Khadayate RS, Sali JV, Patil PP. Acetone vapor sensing properties of screen printed WO_3 thick films. Talanta. 2007;**72**(3):1077–1081

[14] Xu C, Kurokawa J, Miura N, Yamazoe N. Stabilization of SnO_2 ultrafine particles by additives. Journal of Materials Science. 1992;**27**(4):963–971

[15] Dean JA. Lange's Handbook of Chemistry. 12th ed. New York: McGraw Hill; 1979. pp. 10–54

[16] Yamazoe N, Kurokawa Y, Seiyama T. Effects of additives on semiconductor gas sensors. Sensors and Actuators. 1983;**4**:283–289

[17] Yamazoe N. New approaches for improving semiconductor gas sensors. Sensors and Actuators B: Chemical. 1991;**5**(1–4):7–19

[18] Yamazoe N, Miura N. Some basic aspects of semiconductor gas sensors. Chemical Sensor Technology. 1992;**4**:19–42

Section 4

State-of-the-Art Chemical Sensors

Chapter 7

Managing H_2O Cross-Sensitivity Using Composite Electrolyte NO_x Sensors

Erica Perry Murray, Khawlah Kharashi and Kehinde Adedeji

Additional information is available at the end of the chapter

http://dx.doi.org/10.5772/67827

Abstract

NO_x sensors composed of partially stabilized zirconia (PSZ), fully stabilized zirconia (FSZ), and PSZ–FSZ composite electrolytes were investigated using impedance spectroscopy under dry and humidified gas conditions. The impedance data were used to interpret the electrochemical behavior of the various sensors as the water concentration in the gas stream varied. The sensors were operated in the presence of 0–100 ppm NO with 1–18% O_2 and 3–10% H_2O with N_2 as the balance gas. The operating temperature of the sensors ranged from 600 to 700°C. The impedance response for sensors containing ≥ 50 vol% PSZ slightly decreased under humidified gas conditions, in comparison to dry gas conditions; whereas, a significant increase in impedance occurred for sensor largely containing FSZ. This indicated water cross-sensitivity was substantial at FSZ-based sensors. The microstructural properties, NO_x sensitivity, oxygen partial pressure and temperature dependence, as well as the response time of the sensors composed of the various electrolytes were characterized in order to interpret the electrochemical response with respect to water cross-sensitivity. Analysis of the data indicated that sensors composed of a PSZ–FSZ composite electrolyte with 50 vol% PSZ were more suitable for detecting NO_x while limiting water cross-sensitivity.

Keywords: NO_x sensors, porous zirconia, impedancemetric gas sensing, partially stabilized zirconia, YSZ

1. Introduction

Advancements in diesel engine technology and emissions regulations in various countries are driving the need for NO_x sensors capable of greater sensitivity, selectivity, and accuracy. Zirconia-based NO_x sensors are favored for their stability and electrochemical performance under stringent exhaust conditions. Various research studies have achieved substantial NO_x sensitivity at sensors utilizing novel porous electrolytes, as an alternative to the conventional dense electrolyte microstructure [1–3]. NO_x sensors based on the conventional architecture commonly have porous precious metal electrodes to support exhaust gas transport to the electrode/electrolyte interface where NO_x sensing reactions take place. Unfortunately, heterogeneous catalysis can occur within the porous electrodes, thereby reducing the amount of NO_x gas arriving at the interface. This becomes a significant problem when it is necessary to detect low concentrations (i.e., <10 ppm) of NO_x as sensor accuracy is compromised. Heterogeneous catalysis is negligible at NO_x sensors using a porous zirconia-based electrolyte with dense electrodes. The porous electrolyte does not appear to interfere with the exhaust gas during transport to the electrode/electrolyte interface. Thus, the porous zirconia electrolyte enables greater NO_x sensor sensitivity and accuracy. However, cross-sensitivity to interfering exhaust gases, particularly water vapor, adversely affects the selectivity of the device. Sensor operation at specific frequencies appears to be an effective measure for addressing cross-sensitivity to oxygen [1]. Yet, this approach has not been effective for limiting cross-sensitivity to water. Greater understanding of the behavior of water during sensor operation is necessary for managing water cross-sensitivity and improving the feasibility of porous zirconia-based NO_x sensors.

Cross-sensitivity in NO_x sensors can result from electrochemical reactions with exhaust gas species that proceed more readily and/or more rapidly than NO_x reactions. Accumulation of various molecules at the electrode/electrolyte interface can also enable undesirable reactions to proceed leading to cross-sensitivity. The impact of water cross-sensitivity has been ambiguous as it can cause the NO_x sensing response to increase or decrease [4–6]. Some studies suggest that the different behavior observed is related to the sensor materials, fabrication methods, and operating conditions. Few studies have reported the NO_x sensing response for various water concentrations. Limited data with varying results has made it difficult to understand and control the impact of water on the sensing behavior of porous electrolyte NO_x sensors.

Recently, the authors found composite electrolyte NO_x sensors with a porous microstructure demonstrated reduced cross-sensitivity to water, in comparison to sensors with a single-phase electrolyte. To further study the potential of composite electrolytes for NO_x sensing, various sensors based on electrolytes containing specific ratios of partially stabilized zirconia (PSZ) and fully stabilized zirconia (FSZ) were evaluated using the impedancemetric technique. The work presented in this chapter describes the electrochemical response, gas sensitivity, rate limiting mechanisms, and sensing response rate due to operation in humidified gas environments for composite electrolyte NO_x sensors.

2. Experimental

Standard ceramic processing techniques were used to fabricate NO_x sensors containing partially-stabilized zirconia (PSZ, 4.7 mol% Y_2O_3–ZrO_2) produced by advanced ceramics and fully stabilized zirconia (FSZ, 8 mol% Y_2O_3–ZrO_2) manufactured by Tosoh as electrolyte materials. There were five electrolyte slurries prepared using different ratios of PSZ and FSZ along with polyvinyl butyral (Butvar B-76) binder and ethanol for the solvent. The slurries were ball milled for 16 hours to form a uniform mixture. A portion of each electrolyte slurry was dried and the powders were pressed uniaxially at 200 MPa into pellets. The three types of PSZ–FSZ composite electrolytes studied contained 25, 50, and 75 vol% PSZ, and are identified as 25 PSZ–75 FSZ, 50 PSZ–50 FSZ, and 75 PSZ–25 FSZ, respectively. Electrolytes composed of single-phase FSZ or single-phase PSZ particles were fabricated for performance comparison purposes. Single-phase PSZ electrolytes containing coarse particles were also fabricated in order to study the impact of particle size on water cross-sensitivity. Gold (Au) electrodes were placed over the electrolytes and coated using the corresponding electrolyte slurry to complete the sensor configuration. The sensors were air dried and then fired at 1050°C for a period of 1 hour. A schematic of the top view and cross section of the sensors is shown in **Figure 1**. The top view shows the position of the sensor electrodes without the electrolyte coating. The cross-sectional view includes the electrolyte coating over the electrodes. There was no distinction observed between the electrolyte coating and pellet as the materials were of the same composition.

The microstructure and morphology of the electrolytes were analyzed using scanning electron microscopy (SEM) and Archimedes method. The electrochemical behavior of the sensors was characterized using a Gamry Reference 600 to perform impedance spectroscopy. Impedance measurements were performed for sensors operating over a temperature range of 600–700°C where the concentration of NO and NO_2 was varied from 0 to 100 ppm in O_2 concentrations of 1–18% with N_2 as the balance. Data were collected for dry and humidified (3–10% H_2O) environments using a standard gas handling system with mass flow controllers that maintained a flow rate of 100 standard cubic centimeters per minute (sccm). The Gamry instrument was configured to apply a signal amplitude of 50 mV over an operating frequency range of 1 Hz–1 MHz. Measurements were collected in triplicate to insure the data were consistent and stable. Equivalent circuit modeling using Gamry EIS300 software was used to acquire a detailed understanding of the electrochemical behavior of the sensors.

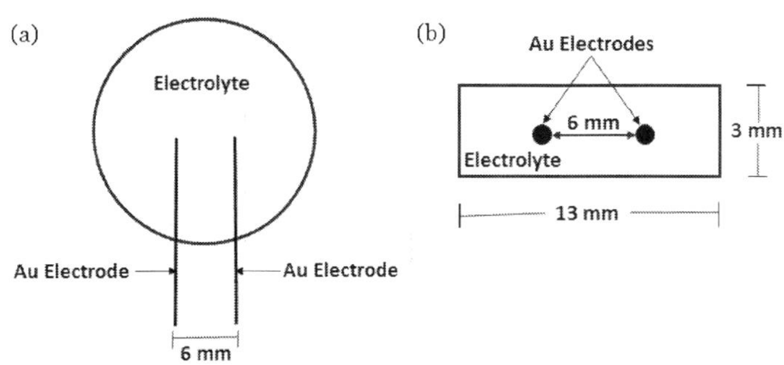

Figure 1. NO_x sensor diagram: (a) top view, and (b) cross-sectional view with embedded Au wire electrodes.

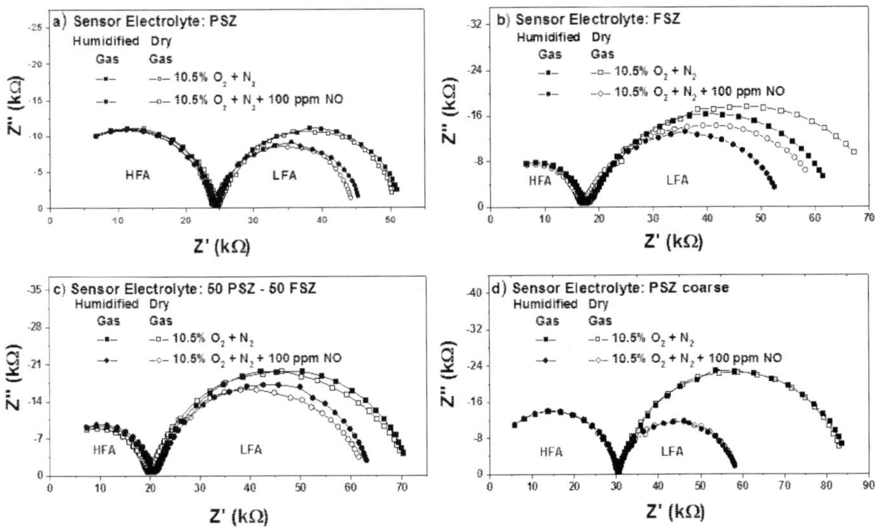

Figure 3. Impedance data collected at 600°C where the sensor electrolyte was: (a) PSZ, (b) 50 PSZ–50 FSZ, (c) FSZ, and (d) PSZ coarse under dry and humidified conditions.

very similar to that for NO. Thermodynamic conversion of NO_2 to NO takes place at elevated temperatures, such that at 600°C approximately 90% of NO_2 converts to NO. For this reason, data using NO gas is presented in this work.

The addition of water to the gas stream caused the low frequency impedance arc to slightly decrease for sensors with a PSZ, 75 PSZ–25 FSZ, and 50 PSZ–50 FSZ electrolyte. Quite the opposite behavior occurred for sensors containing a FSZ and 25 PSZ–75 FSZ electrolyte. The change in the impedance due to water cross-sensitivity is shown for sensors composed of PSZ, FSZ, and 50 PSZ–50 FSZ in **Figure 3a–c**. In other studies, analysis of water adsorption experiments at oxide surfaces have found that molecular water strongly adsorbs onto the surface of Y_2O_3 and surface reactions result in the formation of hydroxyl groups [10, 11]. Furthermore, computational studies indicate dissociation of water molecules is a mechanism for the formation of hydroxyl species at Y_2O_3–ZrO_2 surfaces; and, interfacial reactions between the oxide and hydroxyl groups can enhance oxygen ion conductivity [12]. However, it is also possible for adsorbed water molecules and hydroxyl species to block oxygen adsorption sites along the electrolyte/electrode interface and subsequently hinder interfacial reactions with oxygen [5]. In the present study, the Y_2O_3 content of the sensor electrolyte was approximately 4.7 and 8 mol% for PSZ and FSZ supported sensors, respectively. It is possible that the higher Y_2O_3 content of FSZ allowed greater adsorption of molecular water and hydroxyl groups to take place at sensors containing an FSZ based electrolyte, in comparison to sensors with ≥50 vol% PSZ. This would decrease the available sites for NO and O_2 adsorption. For such as case, triple-phase-boundary reactions requiring NO and O_2 would be limited, thereby, causing the sensor impedance to increase in the presence of water, as shown in **Figure 3b**. The slight decrease in impedance for PSZ sensors likely occurred due to enhanced oxygen ion conductivity. The lower Y_2O_3 content of

PSZ possibly resulted in less molecular water coverage such that sufficient adsorption of oxygen was able to take place. The adsorbed oxygen along with resulting hydroxyl species participated in reactions that produced oxygen ions, which enabled triple-phase-boundary reactions to more readily proceed at PSZ-based sensors.

As mentioned previously, the microstructure of the sensor electrolyte affects the number of particles and reaction sites along the triple-phase-boundary where NO_x reactions take place. The SEM images in **Figure 2** indicate the sensor electrolyte microstructure differed for each sensor type. To determine the impact of the electrolyte microstructure on water cross-sensitivity, impedance measurements of sensors composed of coarse PSZ particles were collected. The impedance data indicated the electrochemical response for the sensors containing the course PSZ electrolyte particles did not significantly change when humidified gases were present, as shown if **Figure 3d**. Since similar behavior was observed for sensors composed of smaller PSZ particles it is unlikely that the microstructure of the electrolyte influenced the NO_x sensor response during humidified gas conditions.

5. Equivalent circuit analysis

To further interpret the electrochemical response of the sensors equivalent circuit analysis was carried out. The impedance arcs in **Figure 3** were simulated using an equivalent circuit model consisting of resistors, R_1 and R_2, and constant phase elements, CPE_1 and CPE_2. **Figure 4** shows the equivalent circuit model that was used to analyze the impedance data for each of the sensors in greater depth. The components R_1 and CPE_1 described the resistance and nonideal capacitance behavior of the electrolyte. Components R_2 and CPE_2 corresponded to the interfacial resistance and nonideal capacitance behavior at the electrodes. The constants n_1 and n_2 describe the deviation from ideal capacitance behavior.

The behavior of R_2 and CPE_2 are useful for gaining insight about electrode reactions that influence sensing behavior. It was found that the interfacial resistance, R_2, decreased as the operating temperature of the sensor increase. This was expected as interfacial reactions, such as charge transfer, are able to proceed more readily at higher temperatures. The relationship between the capacitance, C, and constant phase element, CPE_2 is described by the following equation:

$$C = \frac{[(R_2)(CPE_2)]^{1/n}}{R_2} \tag{2}$$

where $n = 1$ describes the behavior of an ideal capacitor. Studies have reported that oxygen coverage at the electrode/electrolyte interface is described by the low frequency arc capacitance [13]. Oxygen accumulation at the electrode/electrolyte interface can hinder NO_x transport to the interface and subsequently limit NO_x sensing reactions. **Figure 5** shows the capacitance associated with sensors composed of PSZ, FSZ, and 50 PSZ–50 FSZ for dry, and humidified gas conditions. The capacitance for sensors with a PSZ electrolyte tended to be an order of magnitude greater than the capacitance of FSZ electrolyte sensors. It is possible that the difference in grain size between the PSZ and FSZ particles contributed to this difference in capacitance [14]. The capacitance of the 50

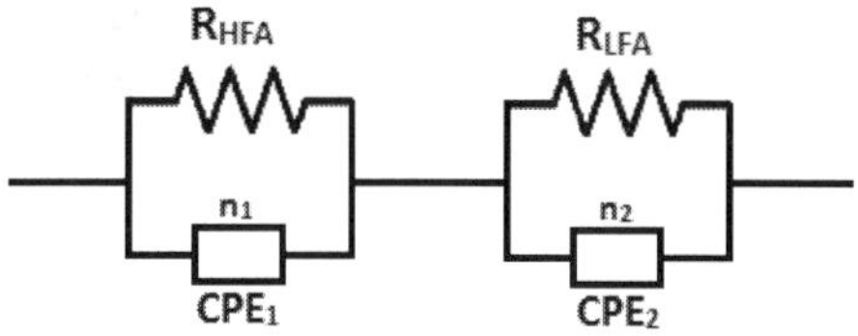

Figure 4. Equivalent circuit used to model the impedance data for PSZ, FSZ and composite PSZ–FSZ based sensors.

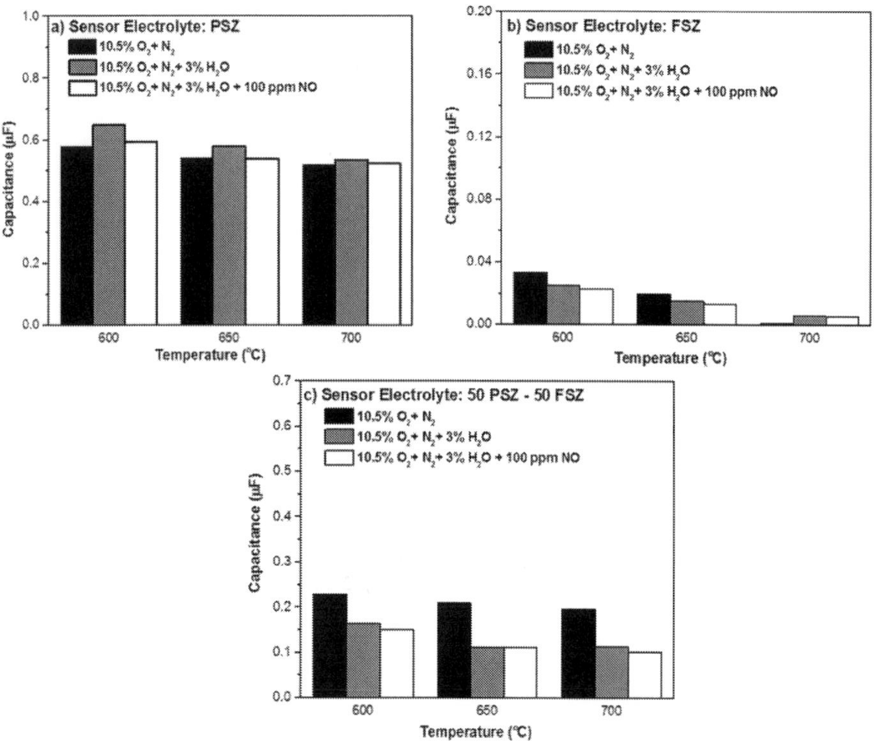

Figure 5. The capacitance of (a) PSZ, (b) FSZ and (c) 50 PSZ–50 FSZ based sensors with for dry and humidified gas conditions with respect to temperature.

PSZ–50 FSZ composite electrolyte sensors was about half of the capacitance calculated for the PSZ-based sensors. Adding 3% water to the gas stream caused a slight increase in the capacitance for the PSZ-based sensors; whereas, a decrease in the capacitance occurred for FSZ electrolyte sensors and especially the 50 PSZ–50 FSZ-based sensors. Adding additional water did not cause further changes in the capacitance. The addition of NO did not significantly alter the capacitance, which suggests NO molecules did not significantly impact oxygen coverage at the interface. The data in **Figure 5** also indicated a strong temperature dependence for capacitance of the FSZ-based sensors, http://dx.doi.org/10.5772/67827

whereas, the PSZ and 50 PSZ–50 FSZ-based sensors displayed a less significant decrease in capacitance as the sensor-operating temperature increased. Overall, it appears the FSZ content in the composite electrolytes was beneficial for by limiting oxygen coverage at the interface such that water and NO molecules had greater opportunity to participate in triple-phase-boundary reactions. The PSZ component of the composite electrolyte helped to limited temperature dependence.

6. Sensor sensitivity

The angular phase component of the impedance, θ, is often more responsive the changes in NO_x concentration, in comparison to other impedance parameters [1, 15]. For this reason, it is frequently used to evaluate NO_x sensor sensitivity according to the following relationship:

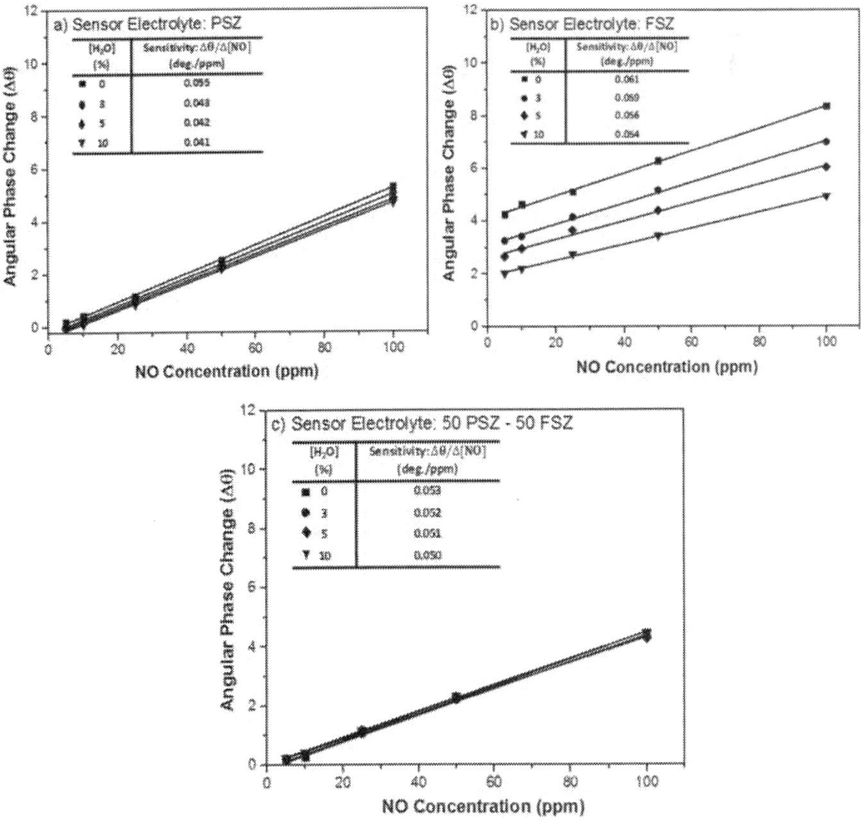

Figure 6. NO sensitivity for sensors with: (a) PSZ, (b) FSZ, and (c) 50 PSZ–50 FSZ electrolyte under dry and humidified gas conditions at 650°C.

$$\theta = arctan\left(\frac{Z''}{Z'}\right) \quad (3)$$

The change in the angular phase, $\Delta\theta$, is given by:

$$\Delta\theta = \theta_{O2} - \theta_{NO} \quad (4)$$

where θ_{O2} is the baseline response when 10.5% O_2 is present with N_2; and, θ_{NO} corresponds to the response when a specific amount of NO gas is added. **Figure 6** shows $\Delta\theta$ with respect to NO and water concentration for the PSZ, FSZ, and 50 PSZ–50 FSZ composite-based sensors at an operating frequency of 40 Hz at 650°C. Data was collected at 40 Hz as the maximum sensitivity was achieved at this operating frequency. The slope of the data, $\Delta\theta/\Delta[NO]$, is defined as the sensor sensitivity in units of degrees/ppm NO. The highest sensitivity was achieved during dry gas conditions for the various sensors. As the water concentration was increased in the gas

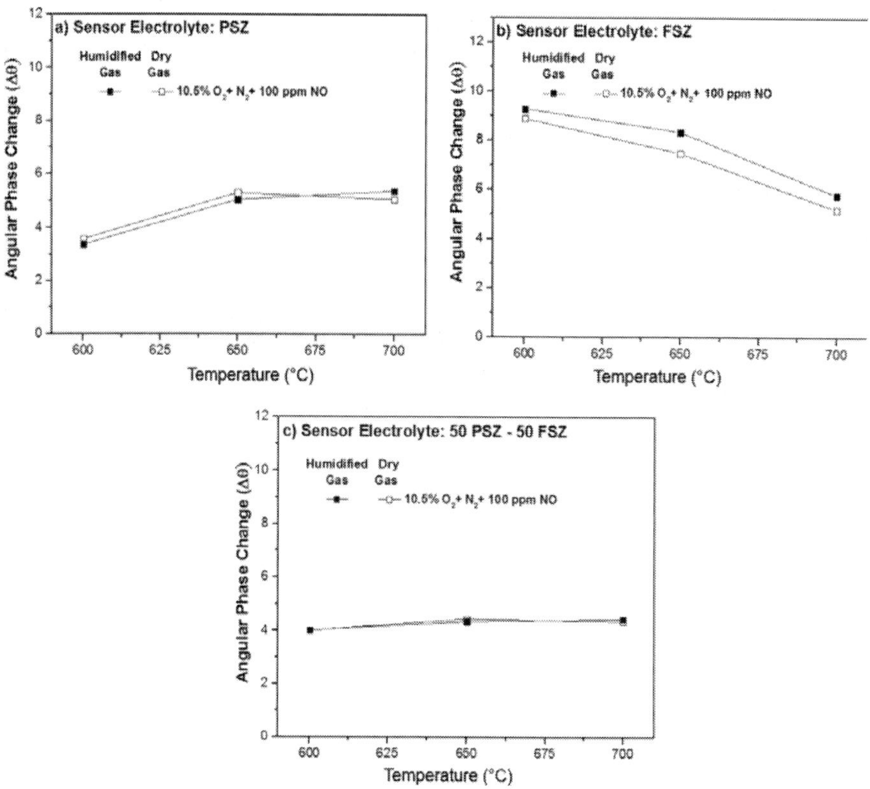

Figure 7. Change in angular phase response with respect to temperature for sensors composed of: (a) PSZ, (b) FSZ, and (c) 50 PSZ–50 FSZ electrolyte.

stream the sensitivity of the sensors decreased. However, the 50 PSZ–50 FSZ composite based sensors demonstrated the greatest tolerance to changes in the gas humidity as the sensitivity decreased very slightly. The sensors with an FSZ electrolyte had a higher sensitivity than sensors composed of PSZ, but the sensitivity of the FSZ based sensors decreased most significantly with increasing water in the gas stream. Although the FSZ electrolyte sensors provide higher sensitivity, the 50 PSZ–50 FSZ electrolyte contributes to a more reliable sensor.

In application the temperature of the sensor can fluctuate due to changes driving conditions. Thus, a stable NO_x sensing response with respect to temperature is necessary in order to maintain accuracy. In the present study, the temperature dependence of the various sensors was observed by comparing $\Delta\theta$ with respect to the sensor-operating temperature, as shown in **Figure 7**. The sensing response of the FSZ-based sensors decreased with increasing operating temperature, as shown in **Figure 7b**. However, the PSZ and especially the 50 PSZ–50 FSZ based sensors were less dependent on temperature, as shown in **Figure 7a** and **c**. This data corresponded with the capacitance data presented in **Figure 5** where the capacitance of the sensors with an FSZ electrolyte also had a strong temperature dependence, in comparison to sensors with a PSZ and PSZ–FSZ electrolyte. Based on the data shown in **Figures 6** and 7, sensors with the 50 PSZ–50 FSZ electrolyte were less prone to water cross-sensitivity and more tolerant to changes in the operating temperature.

7. Rate limiting mechanisms and activation energies

The gases traveling through the porous electrolyte and reacting at the triple-phase-boundary undergo various reaction steps including adsorption, dissociation, diffusion, charge transfer, and oxygen ion transport (not necessarily in that order). The porous microstructure creates specific pathways for gas and ionic transport. The electrolyte material and microstructure can impact how readily gas transport and related reactions occur. A common approach to interpreting rate-limiting mechanisms that impact gas transport and associated reactions is to evaluate the oxygen partial pressure (P_{O2}) dependence, which is described by the following power law relationship:

$$(P_{O2})^m \propto R_{LFA} \tag{5}$$

The power law exponent, m, is used to interpret specific rate-limiting mechanisms, and R_{LFA} is the low frequency arc previously discussed. The P_{O2} dependence of the PSZ, FSZ, and 50 PSZ–50 FSZ based sensors was determined by measuring the impedance response at oxygen concentrations ranging from 1 to 18% for dry and humidified gas conditions with and without NO. A plot of $\log(R_{LFA})$ versus $\log(P_{O2})$ was generated for the various gas conditions in order to determine the exponent m. **Table 2** provides the m values determined for the sensors with a PSZ, FSZ, and 50 PSZ–50 FSZ electrolyte. Dissociative adsorption is the rate-limiting mechanism commonly associated with at P_{O2} dependence where $m = 0.5$ [11, 16]. The data in the table indicated under dry gas conditions $m \approx -0.5$ for the PSZ-based sensors, which suggested dissociative adsorption was rate-limiting. Adding water to the gas stream without NO present decreased m to -0.43 suggesting an additional rate-limiting mechanism, possibly charge transfer, became involved.

	Power law exponent, m			
	NO (ppm)	PSZ	50 PSZ–50 FSZ	FSZ
Dry	0	−0.494	−0.620	−0.412
Gas	100	−0.465	−0.591	−0.413
Humidified	0	−0.427	−0.626	−0.471
Gas	100	−0.512	−0.553	−0.472

Table 2. Power law exponent for sensors composed of the various electrolytes.

The charge transfer becomes a dominant rate-limiting step when $(P_{O2})^{-0.25} \propto R_{LFA}$. In the presence of NO and water, the PSZ-based sensors were apparently limited by dissociative adsorption as a $(P_{O2})^{-0.51}$ dependence was observed. A slightly lower P_{O2} dependence was determined for the FSZ-based sensors. However, in the presence of both water and NO the dissociative adsorption seemed to be the dominant rate-limiting mechanism for FSZ-based sensors. Sensors with the composite 50 PSZ–50 FSZ electrolyte had a stronger P_{O2} dependence, in comparison to the other sensors evaluated. The exponent, *m* tended to be about −0.6, which suggested gas diffusion, may have participated as an additional rate-limiting mechanism. A $(P_{O2})^{-1}$ dependence has been associated with gas diffusion limitations. The porosity of the 50 PSZ–50 FSZ sensor was about 47%, which is expected to provide sufficient gas diffusion. However, it is possible that the composite microstructure resulted in tortuous pathways that interfered with gas diffusion. As a result the dissociative adsorption and gas diffusion could serve as rate-limiting mechanisms for sensors with a composite electrolyte.

The activation energy associated with the various sensors also aids interpretation of how readily sensor reactions are able to proceed. The activation energy, E_a, of the sensors for operating temperatures ranging from 600 to 700°C was determined using the Arrhenius equation the given below:

$$ln\left(\frac{1}{R_{LFA}}\right) = \left(\frac{E_a}{R}\right)\left(\frac{1}{T}\right) + \ln(A_o) \quad (6)$$

where R_{LFA} is the diameter of the low frequency impedance arc, R represents the ideal gas constant, T is temperature, and A_o is the y-intercept. **Table 3** shows the activation energy

	Activation energy (eV)					
	NO (PPM)	PSZ	75 PSZ–25 FSZ	50 PSZ–50 FSZ	25 PSZ–75 FSZ	FSZ
Dry	0	1.067	1.119	1.148	1.150	1.230
Gas	100	1.017	1.101	1.122	1.125	1.205
Humidified	0	1.084	1.110	1.122	1.133	1.135
Gas	100	1.035	1.043	1.099	1.081	1.082

Table 3. Activation energy values for the various sensors.

http://dx.doi.org/10.5772/67827

calculated for each sensor type for dry and humidified conditions with and without NO present. The activation energy was the lowest for the single-phase PSZ-based sensors. The addition of water resulted in an increase in activation energy for the PSZ-based sensors and a decrease in the composite and FSZ-based sensors. Adding FSZ to the sensor electrolyte caused the activation energy to increase. This trend was observed for each of the various gas compositions measured and likely relates to the electrolyte microstructure. Other studies have reported that the activation energy in porous electrolyte NO_x sensors tends to increase with increasing electrolyte porosity [3]. **Table 1** shows the porosity of the electrolytes increased as the FSZ content in the sensor electrolyte increased. The increase in porosity causes particle contact along the electrode/electrolyte interface to decrease resulting in fewer sites for reactions to take place, thereby, resulting in an increase in the activation energy.

8. Sensor response time and stability

Time-based measurements for the angular phase response, θ, were collected for sensors with a PSZ, FSZ, and composite PSZ–FSZ electrolyte. **Figure 8** shows the typical time-based response for sensors with a 50 PSZ–50 FSZ electrolyte as the NO composition was varied over 0–100 ppm. The dry and humidified data overlap as the 50 PSZ–50 FSZ electrolyte enables the sensor to be less prone to water cross-sensitivity. A baseline shift was observed for the time-based data collected for sensors composed the other electrolytes studied. The response time for each sensor was evaluated based on the τ_{90} response, which is the time required for the sensor to achieve 90%

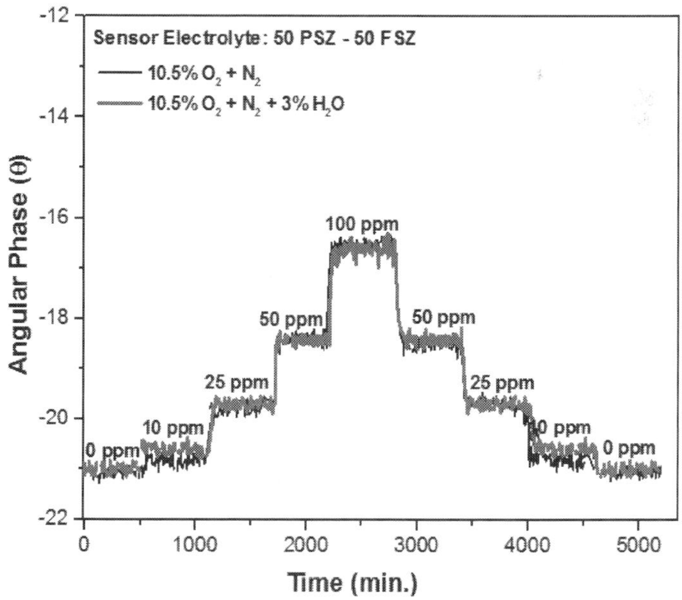

Figure 8. Angular phase response with respect to time for sensors composed of a 50 PSZ–50 FSZ electrolyte.

of the steady state response after a step change in gas concentration. **Table 4** shows the τ_{90} response for the various sensors. The sensors containing single-phase PSZ were the slowest as $\tau_{90} = 16$ s, which possibly relates to the lower ionic conductivity of the electrolyte, in comparison to single-phase FSZ-based sensors where $\tau_{90} = 5$ s. As seen in **Table 4**, the sensor response time decreased resulting in a faster sensor as the FSZ concentration increased.

The stability of a 50 PSZ–50 FSZ based sensors was evaluated for over 150 hours at 650°C and an operating frequency of 40 Hz with 3% water present. **Figure 9** shows sensor drift was negligible for both the baseline response where θ remained about −17.25°, and the sensor response with humidified gas where θ remained nearly constant at about −16.25°. This data suggests that the 50 PSZ–50 FSZ electrolyte is capable of providing a very stable sensing response. Further evaluation is necessary to determine the extent of the 50 PSZ–50 FSZ based sensor stability over longer time intervals.

	Time constant – τ_{90} (s)				
Sensor electrolyte	**PSZ**	**75 PSZ–25 FSZ**	**50 PSZ–50 FSZ**	**25 PSZ–75 FSZ**	**FSZ**
Dry gas	16	14	10	7	5
Humidified gas	14	12	11	5	4

Table 4. The τ_{90} response of the sensors composed of the various electrolytes.

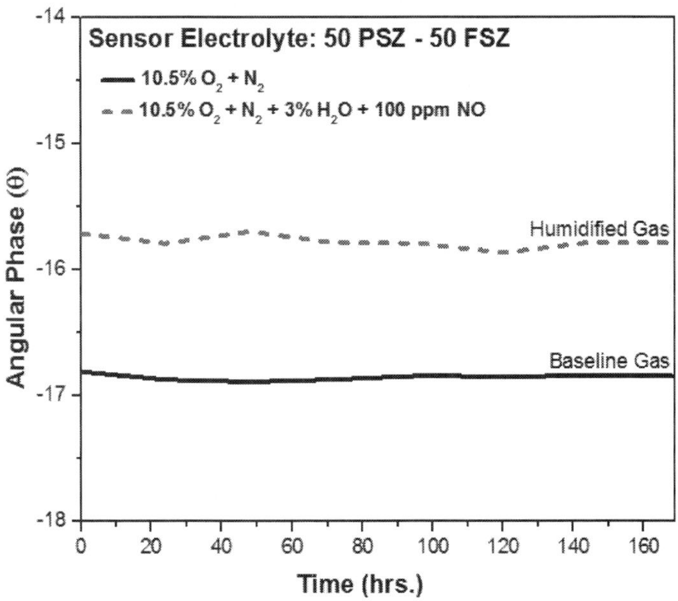

Figure 9. Stability of a 50 PSZ–50 FSZ sensor over several hours of operation.

9. Conclusions

Impedance spectroscopy was used to interpret the electrochemical response of NO_x sensors composed of PSZ, FSZ, and PSZ–FSZ composite electrolytes during operation under dry and humidified gas conditions. Analysis of the electrochemical response of the 50 PSZ–50 FSZ based sensors indicated PSZ contributed to lower water cross-sensitivity, while FSZ promoted NO_x sensitivity. It was also determined that the electrolyte microstructure influenced sensor sensitivity, but did not impact water cross-sensitivity. Dissociative adsorption was considered to be the dominant rate-limiting mechanism for each of the sensors. Adding water to the gas stream did not appear to significantly alter the rate-limiting mechanism. The response of the FSZ-based sensors was strongly depended upon the operating temperature, whereas the response of sensors containing PSZ, and particularly the 50 PSZ–50 FSZ, was less dependent on temperature. The response time of the sensors decreased as the concentration of FSZ increased within the sensor electrolyte indicating the FSZ contributed to a rapid sensor response rate. Overall, sensors composed of the 50 PSZ–50 FSZ composite electrolyte demonstrated significant sensitivity to NO and low cross-sensitivity to water with negligible temperature dependence.

Acknowledgements

The authors thank Mr. Robert Novak and Dr. Jaco Visser of Ford Motor Company for providing meaningful discussions that contributed to this work. Funding for this work was provided by the National Science Foundation under the Ceramics Program (DMR-1410670).

Author details

Erica Perry Murray*, Khawlah Kharashi and Kehinde Adedeji

*Address all correspondence to: emurray@latech.edu

Louisiana Tech University, Ruston, Louisiana, USA

References

[1] Martin LP, Woo LY, Glass RS. Impedancemetric NO_x sensing using YSZ electrolyte and YSZ/Cr_2O_3 composite electrodes. Journal of the Electrochemical Society. 2007;**154**:J97–J104. DOI: 10.1149/1.2430646

[2] Woo LY, Glass RS, Novak RF, Visser JH. Diesel engine dynamometer testing of impedancemetric NO_x sensors. Sensors and Actuators B: Chemical. 2011;**157**:115–121. DOI: 10.1016/j.snb.2011.03.034

[3] Cui L, Han F, Dai W, Murray EP. Influence of microstructure on the sensing behavior of NO_x exhaust gas sensors. Journal of the Electrochemical Society. 2014;**161**:B34–B38. DOI: 10.1149/2.019403jes

[4] Soltis RE, Ding Y, Visser JH, Kubinski DJ. Influence of H_2O on NO_x sensors. ECS Transactions. 2006;**10**:173–178. DOI: 10.1149/1.2357257

[5] Sakai N, Yamaji K, Horita T, Xiong YP, Kishimoto H, Brito ME, Yokokawa H. Erratum to effect of water on electrochemical oxygen reduction at the interface between fluorite-type oxide-ion conductors and various types of electrodes. Solid State Ionics. 2005;**176**:2327–2333. DOI: 10.1016/j.ssi.2005.08.002

[6] Kharashi K, Murray EP. Effect of Al_2O_3 in porous zirconia electrolytes for NO sensing. Journal of the Electrochemical Society. 2016;**163**:B633–B637. DOI: 10.1149/2.0761613jes

[7] Steil MC, Thevenot F, Kleitz M. Densification of Yttria-Stabilized Zirconia Impedance Spectroscopy Analysis. Journal of the Electrochemical Society.1997;**144**:390–398. DOI: 10.1149/1.1837416

[8] MacDonald JR, Barsoukov E. Impedance Spectroscopy: Theory, Experiment and Applications. 2nd ed. New York: Wiley; 2005. DOI: 10.1002/0471716243

[9] Zhang C, Li C, Zhang Z, Ning X, Liao H, Coddet C. Ionic conductivity and its temperature dependence of atmospheric plasma-sprayed Yttria stabilized zirconia electrolyte. Materials Science and Engineering B. 2007;**137**:24–30. DOI: 10.1021/1a9917031

[10] Kuroda Y, Hamano H. Mori T, Yoshikawa Y, Nagao Y. Specific adsorption behavior of water on a Y_2O_3 surface. Langmuir. 2000;**16**:6937–6947. DOI: 10.1016/j.mseb.2006.10.005

[11] Raz S, Sasaki K, Maier J, Riess I. Characterization of adsorbed water layers on Y_2O_3-doped ZrO_2. Solid State Ionics. 2001;**143**:181–204. DOI: 10.11016/S0167-2738(01)00826-8

[12] Chaopradith DT, Scanlon DO, Catlow RA. Adsorption of water on Yttria-stabilized zirconia. The Journal of Physical Chemistry C. 2015;**19**:22526–22533. DOI: 10.1021/acs.jpcc.5b06825

[13] Woo L, Martin LP, Glass RS, Wensheng W, Sukwon J, Gorte RJ, Murray EP, Novak RF, Visser JH. Effect of electrode composition and microstructure on impedancemetric nitric oxide sensors based on YSZ electrolyte. Journal of the Electrochemical Society. 2008;**155**:J32–J40. DOI: 10.1149/1.2804766

[14] Guo X, Maier J. Grain boundary blocking effect in zirconia: a Schottky barrier analysis. Journal of the Electrochemical Society. 2001;**148**:E121–E126. DOI: 10.1149/1.1348267

[15] Orazem M, Tribollet B. Electrochemical Impedance Spectroscopy. Hoboken: Wiley; 2008. DOI: 978-0-470-04140-6

[16] Van Herle J. McEvoy A. Thampi KR. Oxygen reduction at porous and dense cathodes for solid oxide fuel cells. Conductivity Measurements Electrochemica Acta. 1994;**29**:1675–1680.